Union Public Library

AFGHANISTAN	ITALY
ARGENTINA	JAMAICA
AUSTRALIA	JAPAN
AUSTRIA	KAZAKHSTAN
BAHRAIN	KENYA
BERMUDA	KUWAIT
BOLIVIA	MEXICO
BOSNIA AND HERZEGOVINA	THE NETHERLANDS
BRAZIL	NEW ZEALAND
CANADA	NIGERIA
CHILE	NORTH KOREA
CHINA	NORWAY
COSTA RICA	PAKISTAN
CROATIA	PERU
CUBA	THE PHILIPPINES
EGYPT	PUERTO RICO
ENGLAND	RUSSIA
ETHIOPIA	SAUDI ARABIA
FRANCE	SCOTLAND
REPUBLIC OF GEORGIA	SENEGAL
GERMANY	SOUTH AFRICA
GHANA	SOUTH KOREA
GREECE	SPAIN
GUATEMALA	SWEDEN
ICELAND	TAIWAN
INDIA	TURKEY
INDONESIA	UKRAINE
IRAN	UZBEKISTAN
IRAQ	VENEZUELA
IRELAND	VIETNAM
ISRAEL	

Puerto Rico

José Javier López
Minnesota State University

Series Consulting Editor
Charles F. Gritzner
South Dakota State University

Frontispiece: Flag of Puerto Rico

Cover: Colonial architecture along a cobblestoned street in Old San Juan.

Puerto Rico

Copyright © 2006 by Infobase Publishing

All rights reserved. No part of this book may be reproduced or utilized in any form or by any means, electronic or mechanical, including photocopying, recording, or by any information storage or retrieval systems, without permission in writing from the publisher. For information contact:

Chelsea House
An imprint of Infobase Publishing
132 West 31st Street
New York NY 10001

Library of Congress Cataloging-in-Publication Data

Lopez, Jose Javier, 1969-
  Puerto Rico / Jose Javier Lopez.
    p. cm. — (Modern world nations)
  Includes bibliographical references and index.
  ISBN 0-7910-8798-0 (hard cover)
  1. Puerto Rico—Juvenile literature. 2. Puerto Rico—Geography—Juvenile literature. I. Title. II. Series.
  F1958.3.L67 2005
  972.95—dc22   2005028216

Chelsea House books are available at special discounts when purchased in bulk quantities for businesses, associations, institutions, or sales promotions. Please call our Special Sales Department in New York at (212) 967-8800 or (800) 322-8755.

You can find Chelsea House on the World Wide Web at http://www.chelseahouse.com

Text design by Takeshi Takahashi
Cover design by Keith Trego

Printed in the United States of America

Bang 21C 10 9 8 7 6 5 4 3 2 1

This book is printed on acid-free paper.

All links, web addresses, and Internet search terms were checked and verified to be correct at the time of publication. Because of the dynamic nature of the web, some addresses and links may have changed since publication and may no longer be valid.

Table of Contents

1	Introducing Puerto Rico	8
2	Physical Geography	14
3	Historical Geography	32
4	People and Culture	46
5	Political Geography	56
6	Economic Geography	71
7	Living in Puerto Rico Today	81
8	Puerto Rico Looks Ahead	88
	Facts at a Glance	92
	History at a Glance	96
	Further Reading	100
	Websites	101
	Index	102

Puerto Rico

1

Introducing Puerto Rico

As tourists walk along the ocean in Puerto Rico's Old San Juan, few fail to admire the sixteenth-century stone walls that rise magnificently above the azure waters of the Atlantic Ocean. Narrow cobblestone streets lead the visitor into the charm of an old colonial city, where rows of unique shops and restored houses recall a bygone era. Visiting Old San Juan is like traveling into the past, to a time when its military fortifications helped to protect the Spanish Empire that spread across Middle and South America.

Not far from the colonial jewel that is Old San Juan is San Juan itself, Puerto Rico's modern capital city and port, crowded and vibrant. Here are international banks, transnational corporations, busy hospitals and schools, and modern factories—which serve and employ the population of Puerto Rico's largest city. Then, a mere 25 miles (40 kilometers) southeast of pulsating San Juan exists an oasis of serenity and luxurious vegetation that ecotourists consider to be a gem of the Caribbean. This place of cool rivers and waterfalls and exquisite floral beauty is the Caribbean National Forest, where thousands of Puerto Ricans and tourists seek peace and harmony with nature.

Resolutely Hispanic in culture, Puerto Rico has close historical and political ties with the United States, dating to the Spanish-American War of 1898. At first, despite U.S. occupation of Puerto Rico, the island and its peoples were unknown to most Americans. This situation has changed over the years. Today, the island is famous for its sandy beaches, lush rain forests, and dynamic urban areas that make it a popular tourist destination. Inaccurately named the "Rich Port" by the first Spanish settlers, Puerto Rico is not necessarily rich in terms of natural resources, but it is generously supplied with verdant landscapes and progressive, talented, and resilient people. Puerto Rico's 4 million inhabitants have one of the most prosperous economies in all of Latin America.

Puerto Rico, an island only 111 miles long by 39 miles wide (185 by 63 kilometers), is about three times the size of the state of Rhode Island. Despite its relatively small size, it is marked by great diversity and contrast in its landscape, its people, its vegetation, and its productive activities. The flatness and tropical heat of the coastal lowlands is very different from the cool and rugged hills of the interior. The sparsely populated dense forest of the interior's highlands contrasts with the crowded cities of the north coast. The island's climate is tropical, but there are variations within this broad definition. The north coast and the mountainous interior are rainy, but the southern lowlands relatively dry.

Equally diverse are the people of Puerto Rico, with their mainly Hispanic, African, and native Amerindian roots. The culture of Puerto Rico reflects the contributions made by its many different peoples. Afro-Caribbean and Latin American music, dance, theater, and handicrafts are the island's most important artistic forms. Each region of the island has its own

Puerto Rico, the easternmost of the Greater Antilles group of islands, is known for its tropical climate and sandy beaches.

cultural identity, but there are common themes as well. For example, Puerto Ricans from the countryside enjoy cockfights. In rural areas, a cockfight is bound to attract energized crowds from all over the island. The most common and pervasive of all cultural inputs now come from the North American mainland. The United States is represented everywhere by films, music, styles of dress, and products on supermarket shelves. Nevertheless, the people of Puerto Rico proudly retain their own heritage: They study English but speak only Spanish and wherever possible resist the intrusion of "Yankee" culture.

Puerto Rico has many heroes and celebrities, past and present, whose names appear frequently in international newspapers. The island has produced fine singers, including Jose Feliciano and Ricky Martin. Movie stars of stature include Jose Ferrer, Rita Moreno, Benicio del Toro, and Raul Julia. Puerto Rico has enriched the world of sports with excellent baseball players such as Roberto Clemente, Ivan Rodriguez, and Roberto Alomar. Other Puerto Rican sport heroes include golfer Chi Chi Rodriguez, tennis player Gigi Fernandez, and basketball star Piculin Ortiz. Puerto Ricans love horse races, and the island has produced a number of outstanding jockeys. Among the more famous Puerto Rican jockeys is Angel "Junior" Cordero.

In many ways, Puerto Rico functions as the hub of the Caribbean. The island is strategically located midway between the southeastern part of the United States and South America's northern coast. The channels that separate Puerto Rico from Hispaniola (which includes the countries of Haiti and the Dominican Republic) and the Virgin Islands give ships coming from Europe and Northern Africa access to Venezuela, Colombia, and the Panama Canal. In essence, Puerto Rico links Europe with South America. Every vessel navigating to or from Venezuela and Colombia has to go through the island's territorial waters. During the colonial era, the island's strategic significance was obvious to the Spanish authorities, especially after the establishment of colonies and mines in the northern part of South America. San Juan was among the chief provisioning ports of the area, for both Caribbean-bound and homeward-bound Spanish vessels.

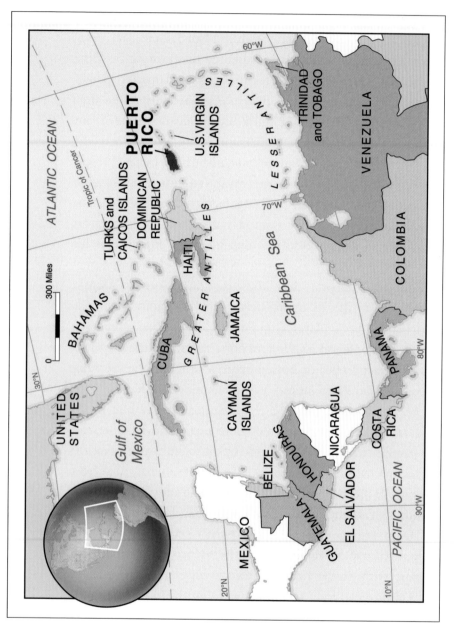

Puerto Rico is strategically located midway between the southeastern part of the United States and South America's northern coast. It is bounded on the north by the Atlantic Ocean, on the east by the Virgin Islands Passage, on the south by the Caribbean Sea, and on the west by the Mona Passage, which separates Puerto Rico from the Dominican Republic.

The strategic importance of the channels or sea passages (also known as sea gates) separating Puerto Rico from other islands grew enormously with the completion of the Panama Canal in 1914. Between 1910 and the 1980s, the defense of northeastern approaches to the Panama Canal rested in large measure on the control of Puerto Rico's sea passages. For this reason, the United States, in order to assure its control of the region, built naval, air force, and army bases on the island. During World War II, when German submarines threatened Caribbean Sea lanes, Puerto Rico's bases played a key role in protecting the area.

Today, the amount of sea trade moving through the waters of Puerto Rico is greater than ever. The island is a major hub of Caribbean commerce and tourism. Puerto Rico's main seaport is among the largest container ports in Latin America in terms of volume of movement. Shippers use Puerto Rico as a distribution center to the rest of the Caribbean. In addition, San Juan is also one the world's busiest cruise-ship ports, with cruise piers that can handle ships carrying up to 4,000 people.

Puerto Rico dominates the air transportation system of the Caribbean. San Juan International Airport is the busiest in the region. This international airport is a popular place for hopping over to other Caribbean destinations. The airport services flights to and from smaller islands in the eastern Caribbean that do not have the capacity that Puerto Rico has to handle thousands of passengers on a daily basis. An air cargo airport in northwestern part of the island has the longest runway in the Caribbean.

In the following chapters, you will discover the varied landscapes and charming people of this tropical Caribbean paradise.

CHAPTER 2

Physical Geography

Puerto Rico is the smallest of the Greater Antilles, the group of large Caribbean islands that includes Cuba, Hispaniola, and Jamaica. A mountainous tropical island with several distinct geographic regions, about three-fourths of Puerto Rico's territory is hilly or mountainous. Some geologists have theorized that millions of years ago the islands of the Caribbean were part of a single landmass. With plate movement came volcanic eruptions and earthquakes, eventually causing the islands to become separate entities. In addition to geologic ties, Puerto Rico shares with other Caribbean islands its historic experiences and human adaptation processes. Like the other islands, Puerto Rico has experienced high rates of deforestation, and its once pristine coastal environments have been degraded by human activity.

Puerto Rico can be divided into three types of landform regions: the interior highlands, the coastal lowlands, and the northern limestone

Physical Geography

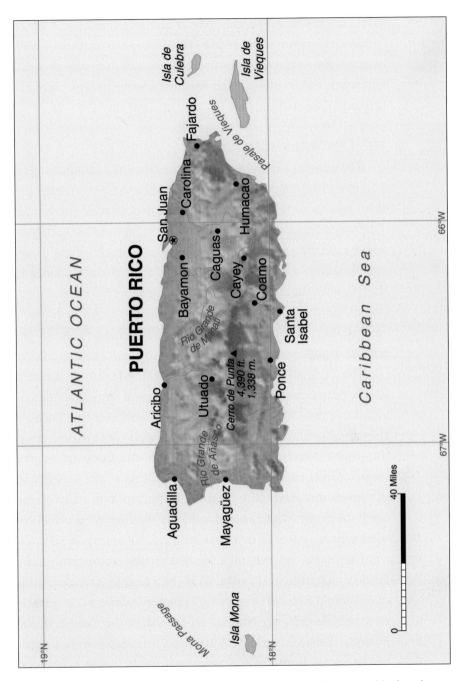

Puerto Rico is a mountainous tropical island with several geographical regions. It can be divided into three types of landform regions: the interior highlands, the coastal lowlands, and the northern limestone karst country.

karst country. Karst topography is unique to limestone areas. In limestone, groundwater can remove large quantities of rock beneath the surface. Caverns are formed during the process, and when these collapse, depressions known as sinkholes are formed at the surface.

THE CENTRAL HIGHLANDS

The central highlands consist of three mountain ranges. The most impressive chain is the Cordillera Central, running from east to west across the south-central part of the island. This range includes Puerto Rico's highest mountains, with four peaks exceeding 4,000 feet (1,220 meters) in elevation. Here, cool streams and isolated, heavily forested areas contrast sharply with the urban industrial character of the coast. Populations are lower, the climate more moderate, and lush green scenery more common. The majority of the island's forests are located in this mountain range.

The crest of the Cordillera Central forms the drainage divide between streams flowing northward into the Atlantic and those flowing southward into the Caribbean Sea. Because these highlands receive significant rainfall, the island's most important dams and reservoirs, which supply water to the larger coastal towns, are located here. Some of the water held in highland reservoirs is used for irrigation, especially of the drier south, whereas some is used for the limited production of electricity.

The Sierra de Luquillo, located in the northeast, is the island's second-largest range. Its highest peaks reach elevations of 3,000 to 3,524 feet (914 to 1,074 meters) above sea level. The range boasts the only tropical rain forest in the United States' National Park system. The Caribbean National Forest, known by the locals as *El Yunque*, is famous for its dense vegetation and heavy precipitation. Moisture-laden air masses from the northeast move across Puerto Rico and strike the northern slopes of the Sierra de Luquillo. This is the rainiest part of the island, with El Yunque receiving an average 200 inches (500 centimeters) of annual precipitation from almost daily showers. Hundreds of different small animal and plant species make the Caribbean National Forest their home. Birds, reptiles, and amphibians dominate the forest's animal life.

Four forest types coexist in the Caribbean National Forest: the *tabonuco, palo colorado, sierra palm,* and dwarf. The tabonuco forest thrives on the very humid lower slopes of the range at elevations below 2,000 feet (600 meters). Tall trees and little light at ground level characterize this forest. Giant ferns also are common. The palo colorado forest, home of the Puerto Rican parrot, occupies elevations of the mountains between 2,000 and 3,000 feet (600 to 900 meters). This mossy forest adjoins the sierra palm forests, which are adapted to wet conditions and to the slippery and unstable soil found on steep slopes. Puerto Rican parrots live in the nearby palo colorado forest but consume the palm fruits. The palm forest extends to elevations around 2,500 feet (750 meters). The mountaintops are covered with dwarf forests. Trees here are short because soils are wet and the prevailing dense fog prevents sunlight to effectively assist in the plant growing process. High winds and lower temperatures are also responsible for the small size of the plants living at these higher elevations.

The attractive waterfalls and panoramic views of the Sierra de Luquillo provide a popular destination in Puerto Rico. Thousands of Puerto Ricans and foreigners visit this mountain range to enjoy the scenery and to swim below its waterfalls. The hiking trails of the national forest offer diverse landscapes and luxuriant tropical views.

The island's third most important range, the Sierra de Cayey, rises up in the southeastern part of the island. Highest elevations in this range are about 2,900 feet (879 meters). A tobacco growing region for many decades, the area's hillsides now tend to be covered with banana trees, plantains, and planted pines.

COASTAL LOWLANDS

Between the mountains and the sea are the coastal lowlands, varying in width from 2 to 8 miles (3 to 13 kilometers). Throughout the island's history, most human activity has occurred on these lowlands. The island's coastal ports have linked Puerto Rico with the rest of the world. In addition, the fertile soils of the coastal plain have favored agriculture, and the flat terrain has provided a favorable setting for industrial and urban growth. Most Puerto Rican towns and cities are located on the coastal lowlands.

The northern lowlands are especially favored. Soils are fertile and water is plentiful. For these reasons, the north early became the island's chief agriculture and livestock-ranching region. These activities played a key role in the island's economy from the 1500s until the first half of the twentieth century. Today, the northern lowlands support the majority of the island's population and its most important industrial activities. Significant deposits of clay and sand are located near the coast, and these have some industrial uses, but otherwise the region lacks mineral resources. The northern coastal plain has been altered by human activity. Many small lakes and wetlands have been drained to expand the amount of available farmland. Along the coast, some of the mangrove swamps have disappeared as a result of urban expansion.

The southern lowland belt is narrower than the northern coastal plain. In part, these plains were formed at the base of alluvial fans (broad, gently sloping deposits formed from gravel where a fast-moving stream emerges from a narrow canyon onto a broad valley floor). Alluvial fans are common in dry areas, and southern Puerto Rico is much drier than the northern part of the island. Beaches of the southern coast are not as sandy as those of the north. Because of steeper south-facing mountain slopes, streams flowing southward into the Caribbean Sea tend to deposit more gravelly materials at their mouths than do those flowing northward. For historical reasons, and because of its physical geography, the southern coast lacks the urbanism, industry, agriculture, and tourist development that the northern coast has experienced.

Mangrove Swamps

Mangrove refers to the highly adapted plants found in tropical swamps. Mangrove plants live between the sea and the land. At high tide, their roots are fully immersed in salt water, and at low tide they come in contact with fresh water. In Puerto Rico's northeastern coast, this unique ecosystem is found at the mouth of rivers or around coastal lagoons. Numerous plants and small animals live in these mangrove wetlands. Nearly 50 bird species inhabit the island's biggest mangrove forest, located just east of San Juan.

Mangroves are beneficial because they protect coastal areas from erosion and serve as a shelter for wildlife. They also offer protection to low-lying areas of the island when tropical storms approach. Converting carbon dioxide into oxygen, these forests are the "lungs" of Puerto Rico's northeastern lowlands. Significant portions of the forests have been eliminated as a result of urbanization, however, so those that remain are considered some of the island's most valuable natural resources. Puerto Rico has established several natural reserves in the northern coast to protect these important ecosystems.

Mangrove lagoons are famous for the bioluminescence (light) produced by dinoflagellates. These harmless single-celled organisms are found in waters surrounded by mangroves. Whenever they sense movement, these tiny creatures give off light. Unfortunately, this nighttime light show is threatened by the clearing of coastal mangrove forests that surround every known remaining bioluminescent bay in Puerto Rico. Many tourists visit the northeastern mangrove lagoons at night to see this incredible show of underwater lights. Tourism is another reason for preserving the mangrove forests.

Mangrove swamps, such as this one near the southwestern town of La Parguera, provide a rich ecosystem for numerous plants and animals and also protect coastal areas from erosion.

THE KARST COUNTRY

The transition zone between Puerto Rico's northern lowlands and central highlands is a region where the land rises in regular green hillocks, with some rocky spots. These hillocks, known locally as *mogotes,* are the product of a process in which water sinks into limestone and erodes larger and larger basins, known as sinkholes. As erosion proceeds, small peaks remain between the sinkholes. In essence, these rounded peaks are the land that was not eroded. The most representative area of karst topography is in the northwest. Impressive caverns, mogotes, and sinkholes dominate this part of the island. Limestone found in this area is white. Its sediments were formed thousands of years ago by seashells, coral, and bones of marine species.

Dense vegetation tends to cover the karst hillocks or mogotes of northern Puerto Rico, providing an excellent shelter to endangered plant and animal species. Mogotes are not suitable places for building construction because of their steep slopes. Unfortunately, during the past 50 years, many of these karst hillocks have been destroyed, as construction companies have used the limestone for building materials or the hills have been eliminated during the construction of highways and roads.

It is not surprising that Puerto Rico's karst country has become an important ecotourism destination. One important feature of the region is the numerous caverns that have been produced by underground rivers or reservoirs. After thousands of years of interaction between limestone and percolating water, the rock dissolved. This process left a great number of large underground cavities (caves) in northwestern Puerto Rico. Entering these caves can be an awesome experience. Many feature columnar formations that are the product of the interaction between water and the rocks. Stalactites hang from the cave ceiling, stalagmites rise from the cave floor, and columns often form where the stalactites and stalagmites meet. Thousands of bats live inside the caves, sleeping during the day and leaving at dusk to forage.

It is estimated that there are 2,000 caves in Puerto Rico, only about 250 of which have been thoroughly explored and studied. One of the largest cave systems in the Americas is located in the northwestern part of the island. The cave can be reached by entering some of the area's sinkholes. In fact, the Camuy River disappears from sight as it flows into some of the sinkholes and through a series of caves. The largest entrances into this complex cave system have been developed for tourism. Sinkholes in the karst country have a round shape and appear like craters. There are about 1,000 sinkholes in the karst country. One of them houses the 1,300-foot- (396-meter-) wide dish of the world's largest radio telescope.

The mogotes and sinkholes facilitate the movement of rainwater to aquifers (natural underground reservoirs of water). These aquifers supply water to farming communities in northern Puerto Rico. Basically, aquifers are replenished every time rainwater runs down the mogotes' slopes into the sinkholes. Because of the porosity of the land's surface, water percolates easily into the underground reservoir. Sadly, some sinkholes are now being used as dumpsites, thereby polluting the water that enters the aquifer.

CLIMATE

Puerto Rico enjoys a pleasant tropical maritime climate. Although it lies within the Tropics, between latitudes 18°50' and 17°55', it is surrounded by water—the Atlantic Ocean to the north and the Caribbean Sea to the south. This prevents extremes in temperature, and prevailing winds continually bring balmy conditions. The average annual temperature is 79° Fahrenheit (26°C), with winter and summer averages of 76°F (25.5°C) and 82°F (28°C) degrees, respectively.

Puerto Rico's climate is dominated by three factors: the overhead position of the sun, prevailing northeast trade winds, and elevation. When the sun is directly or almost directly overhead, convectional thundershowers occur daily and all over the island. This happens between May and October. But in the dry season, which lasts from November through April, the sun's overhead position is south of the Equator and there is little rainfall. Trade winds bring a pleasant breeze. This milder and drier period is the prime season for tourism.

Trade winds and mountains combine to influence the island's rainfall patterns. The trade winds coming from the northeast, off the Atlantic, are loaded with moisture. As they blow inland over the northern coastal plains, they hold most of their moisture. But as they rise against the slopes of the central mountain ranges, they can no longer contain it, and the moisture drops as rain. Trade winds bring heavy rainfall, as much as 200 inches (500 centimeters) a year, to the northern face of the central mountains. This rain is collected in the northward-flowing rivers that water the coastal plain. Once the trade winds have dropped their moisture, they descend the southern face of the mountains. The southern slopes and the coastal plain are dry by comparison. Rivers that reach the southern coast are short, shallow, and dry for half the year.

Puerto Rico has three distinct climate zones. The northern lowlands have a tropical wet-and-dry type of climate, whereas the central mountains have a tropical rain forest on their northern slope and a tropical semi-arid climate on their southern slope. The southern coastal lowlands vary from tropical wet-and-dry to tropical semi-arid.

Highest temperatures occur on the southern coastal plain. Whereas the northern lowlands report an average annual temperature of 77°F (25°C), the southern plains have an average temperature of 80°F (27°C). The central highlands have an annual average temperature of 74°F (23°C), and on the island's highest peaks, the average is 68°F (20°C). For this reason, many islanders enjoy visiting Puerto Rico's interior to escape from the higher temperatures of the lowlands. One advantage of the highlands is that they offer an excellent climate for the commercial farming that produces flowers, strawberries, and poultry for the island's markets. In the highlands, February is the coolest month and nighttime temperatures can drop to 50°F (10°C). The lowest temperature ever reported was 40°F (4.5°C), in 1911. In 1979, the temperature on the southern coast soared to a scorching 106°F (41°C), the island's record high.

In addition to experiencing the island's highest temperatures, the southern lowlands also have the lowest annual rainfall averages, with some areas receiving less than 40 inches (100 millimeters). The island's semi-arid southwestern corner receives less than 30 inches (76 millimeters) annually. This region resembles part of the American Southwest, with cactus and low thorny bushes covering the land. Here, the government maintains a 1,570-acre (6.35-square-kilometer) forest reserve with excellent examples of tropical dry forest vegetation. The effects of the dry winds, little moisture, and considerable amounts of salt in the environment limit plant growth. Only drought-resistant spiny shrubs and trees survive these conditions. However, the land and marine-animal life of the area is diverse.

Numerous types of birds, reptiles, and crustaceans live in this area. Beaches near this dry forest are among Puerto Rico's most popular recreational places. They are famous for their white sands and turquoise waters, with warm temperatures and sunny days always tempting sunbathers to stay. The lack of heavy rains in this area is responsible for the excellent underwater visibility of the reefs, which are visited by numerous snorkelers and scuba divers.

Annual rainfall on the northern lowlands average 60–70 inches (150–180 centimeters). Puerto Rico's most important rivers are located in this zone. The island's longest river, La Plata, is about 50 miles (80 kilometers) long and originates in the central highlands. These highlands receive 70–100 inches (180–200 centimeters) a year. The Sierra de Luquillo has an annual rainfall average of about 150 inches (380 centimeters), with some parts of the range reporting an average of 200 inches (500 centimeters). For that reason, the island's most impressive rain forest is located there.

HURRICANES

Hurricanes are the most devastating natural events that affect the island of Puerto Rico. These ferocious storms can cause loss of human life and widespread destruction of

In September 1998, Hurricane Georges lashed Puerto Rico with 115 mph winds. The powerful storm left at least 17,000 people homeless and 4 million without power, and caused more than $2 billion worth of damage.

property, both in the countryside and in urban areas. Sometimes, it takes several years for the island and its people to fully recover from a particularly disastrous hurricane. Even though science has made significant progress understanding and analyzing these fierce tropical storms, hurricane paths cannot be predicted with certainty. In September of 1998, Hurricane Georges crossed the island from east to west with merciless winds that struck the territory for long hours.

Powerful winds are responsible for much of the damage that can be seen on the island following a tropical storm. Hurricanes also trigger disastrous floods. Since significant portions of Puerto Rico have suffered deforestation, large areas lack the topsoil and vegetation that can hold the heavy rains accompanying a hurricane. When vegetation is removed, the topsoil washes away. Consequently, the land is no longer able to absorb heavy rainfall. For this reason, swollen rivers and devastating floods are occurring with increasing frequency. Not only has deforestation resulted in flooding, but the elimination of large trees has removed a natural protection against the strong winds.

Since the Spaniards arrived (and began keeping records) some 500 years ago, the island has been hit directly by about 100 hurricanes. Before the mid-1900s, storms often had a devastating impact. The island's economy depended almost exclusively on agriculture, and houses and other buildings were poorly constructed. Entire agricultural fields, farms, and plantations were destroyed, and human casualties were high. In terms of loss of life, the island's most tragic storm, which occurred in 1899, was responsible for the deaths of 3,369 people. During recent decades, however, the number of human casualties has sharply declined. Scientific advances have made it possible to alert people well in advance of a storm's arrival, giving them time to seek safe shelter. In addition, houses are better built and the government provides very safe places for those who live in poorly built structures.

EARTHQUAKES

Puerto Rico, lying in a zone where the movement of tectonic plates can be rapid and complicated, has experienced many earthquakes. Tremors are frequent in and near the island, but they tend to be rather weak in magnitude. A 1981 tremor was felt across the island, whereas another earthquake, in 1985, was felt in the towns of south central Puerto Rico. Some studies suggest that the island can expect a large earthquake every 50 or 60 years. The problem with this type of natural disaster is that it cannot be forecast as well in advance as a tropical storm.

In the past 150 years, Puerto Rico has experienced two devastating earthquakes. On October 11, 1918, an earthquake with a magnitude 7.5 was centered approximately 9 miles (15 kilometers) off the island's northwestern coast. This earthquake produced a tsunami that struck several towns along the island's west coast. Of the 116 people killed by this seismic event, 40 were victims of the tsunami. The main square of the western city of Aguadilla was swamped by a wave nearly 20 feet (6 meters) high. Numerous houses, factories, public buildings, bridges, and other structures suffered severe damage during this earthquake and its resulting tsunami. On November 18, 1867, 20 days after Puerto Rico was devastated by a hurricane, a strong earthquake struck the southeastern coast. It produced a tsunami that ran inland almost 500 feet (150 meters). No loss of life was recorded.

PLANT AND ANIMAL LIFE

When the Spanish began colonizing Puerto Rico during the early 1500s, the island was almost entirely covered by dense vegetation. In addition, animal life was abundant with numerous species of birds, fish, crustaceans, amphibians, reptiles, and some mammals. Human occupation of the island has destroyed much of the native flora and fauna, however. Some species that survived the traumatic effects of European occupation are now listed as endangered species.

The Spanish colonizers imported a wide variety of plants and animals from other continents. Bulls, cows, horses, donkeys, pigs, goats, chickens, and dogs were among the animals brought onto the island. The Spanish also introduced to the island Old World cultivated plants, including rice, plantain, bananas, wheat, sugar cane, coffee, and oranges. Puerto Rico's contemporary diet consists largely of dishes made of rice, peppers, squash, chicken, pork, and beef. Interestingly, Europeans adopted grains and vegetables such as maize (corn), beans, squash, and peppers that were native to the Americas.

When the Spanish arrived, the Puerto Rican lowlands were covered with *palo de pollo* or chicken tree, so-named because the shape of its buttresses resembles the foot of a chicken. This massive tree has large leaves and can grow up to 130 feet (40 meters) tall, with a diameter of three feet (1 meter). The palo de pollo tree thrived along rivers and creeks and in swampy areas. Another important tree found on the island was the mighty *ceiba,* which can reach heights of 160 feet (49 meters). Amerindians built their canoes from the trunks of this huge tree. The higuera tree was very important to island natives before the arrival of Europeans. Maracas were made from its fruit. This tree is common throughout Puerto Rico. Indians also used the higuera's fruit to make bowls and cups by cutting the dried shell in half. Some people still use these utensils.

A Caribbean tree commonly used by the Spanish colonists for construction because of the quality of its wood was the *ausubo*. This tree, its wood used for beams in the historic buildings of Old San Juan, is still found on the island. One tree common on the island, but non-native, is the *flamboyant* or flame tree. Its bright red blossoms make it the most colorful tree of the Puerto Rican countryside.

Among the fauna native to Puerto Rico, the most famous— but at the same time the closest to extinction—is the Puerto Rican parrot. Another important bird of the island is the *pitirre*. The pitirre is a small, fast, brave bird that fights to win or to get the respect of the island's local eagle, the Antillean red-tailed hawk, or *guaraguao*. The island's bird life is very diverse. Nightingales, doves, sparrows, and the Puerto Rican stripe-headed tanagers are common throughout the island's interior. Seagulls and pelicans are widespread in the coastal areas, and in some isolated parts of the island it is possible to find owls and falcons.

The manatee is the largest of the island's mammals. This aquatic mammal lives in the delta areas of the island's larger rivers. The eastern coastal area is the manatees' preferred location. This animal is considered endangered in the northeastern part of the Caribbean. Some manatees can be found in limited parts of Puerto Rico's south and western coast. These mammals prefer to live in shallow waters and, because they are herbivorous, are found mostly in areas where aquatic plants are abundant. The fertility rates of the manatee are low, producing only one newborn every three to five years. Young manatees stay with their mother for periods of up to four years in order to receive protection and to learn which places are suitable for sustenance. Even though protected by law, these friendly aquatic mammals face numerous problems that cast doubt on their survival. Many of the areas where manatees normally live have been polluted or even destroyed by the building of harbors. The accumulation of sediments in the habitat of these mammals also threatens their existence. In addition, numerous manatees have been hurt or killed by the propellers of boats that have invaded their zones.

Inland, the *jutia* used to be an important land mammal. A rodent, similar in size to a rabbit, the jutia has a broad head and small ears. Although common in Cuba, Jamaica, and Hispaniola, scientists fear that the jutia may have become extinct in Puerto Rico. Bats are the most numerous mammals living in Puerto Rico.

Turtles, lizards, iguanas, and snakes are representatives of Puerto Rico's reptile population. The most famous Puerto Rican amphibian is a small tree frog known as *coqui*. The coqui is a very popular animal throughout the island because of its charming "song," which can be heard in late afternoons and evenings. Its name comes from the mating call made by the male of the species, sounding like "ko-kee . . . ko-kee."

The most common salt water fish in Puerto Rican waters are the red snapper, blue marlin, king fish, barracuda, grouper, and robalo. Coral reefs can be found off most coasts. These reefs sustain small island fisheries, shelter the shoreline from

After the sun sets, the male coqui serenades the people of Puerto Rico with its charming song. Known as "little frog," the coqui gets its name from its characteristic two-note call: "ko-kee, ko-kee."

erosion, create sandy beaches, and represent perhaps the most important coastal resource of the island. The best-formed reefs are in the shallow waters along the drier southern part of the island, where few rivers deposit sediments on coral formations. Today, some coral reefs are being severely damaged by the dropping and dragging of fish traps and anchors.

MINERALS

Puerto Rico has some deposits of copper, gold, silver, manganese, and iron. The island's resources also include some molybdenum, cobalt, chrome, and nickel. Large-scale mining for these minerals has not occurred, though. Environmentalists, some government agencies, and organizations that favor the conservation of nature have vehemently opposed mining projects. They fear that such projects in a small and heavily populated island would compromise the environmental quality of the territory and the health of its inhabitants. These nature conservation efforts have successfully halted numerous attempts to develop mining complexes in the island.

Another reason these minerals have not been extracted is because of the small size of many deposits. Copper is the only mineral found in relatively large quantities, but mining it could lead to an ecological disaster. Some iron mining took place in the east central part of the island when several American companies operated mines between 1898 and 1953.

Puerto Rico's nonmetallic minerals are quite abundant and diverse. The island's construction companies have used rocks such as granite, limestone, marble, sands, and clay extensively. A significant proportion of Puerto Rico's blue-collar labor force is employed in the extraction and use of these raw materials.

3

Historical Geography

The European explorer and navigator Christopher Columbus caught sight of Puerto Rico on November 19, 1493. When his Spanish crew landed on this island, they were met by Amerindians known as the Tainos. These natives were peaceful farmers and fishermen who called their island *Borinquen,* which means "land of the brave lord." Many of the possessions used by these Amerindians have been found in the northern part of South America. Some anthropologists, therefore, believe that the Tainos came originally from Venezuela. Tainos had large canoes that permitted them to travel and trade widely within the Caribbean basin. The Spanish were quite impressed with the navigational abilities of these native peoples and the speed of their movement across water in their dugouts.

Initially, the Tainos welcomed the Spanish explorers, and the early stages of European occupation were peaceful. In fact, from

1493 until 1508, very little contact occurred between the native people and foreign intruders. The Spanish attention was focused elsewhere, as they explored and colonized other parts of the region. During the early years, only on a few occasions did Spanish sailors land in Puerto Rico. When they did visit the island, they were seeking trade for food, and the encounters were usually welcoming. In 1508, however, conditions changed for the worse as Spanish conquerors arrived to colonize the island. They had found gold on the island. Their lust for this metal severely tarnished their relationship with the Taino people.

THE SPANISH CONQUEST

Juan Ponce de Leon was given the task of colonizing Puerto Rico. At the outset, he participated in a ceremony that was supposed to establish an agreement of friendship and alliance with some of the island's Indian communities. Ponce de Leon's priorities did not include the well-being of the Taino population, however. He was more interested in establishing gold mines in the new colony. In order to support the mining operations and the new colonists, farms were established. Spanish fortune hunters living in the gold districts violently forced the Indians into mining this precious metal. During the initial phases of mining, Puerto Rico produced as much gold as Hispaniola and more than Cuba. In order to keep up the production of gold, the Spaniards established a system that obligated the native populations to supply workers. Not surprisingly, under this system most labor came from Indians who were considered rebellious.

Another method of converting natives into miners was the *encomienda* system. Under this system, a Spanish colonist would be given a grant of land. In return for the grant, he must agree to provide the Tainos with food, shelter, a Christian education, and "humanitarian treatment" under supervision of the authorities. Nevertheless, many colonists failed to obey the system's rules. Because of their greediness and desire for more gold, they often disregarded the welfare of the natives. Colonists involved in this system were supposed to pay the Indians for their work, but they often failed to do so. Interestingly, the colonists discovered that the Taino society was not familiar with the concept of money and salary. Spaniards took advantage of this ignorance and did not compensate the Indians for their hard work. Many natives died as a result of the abusive treatment they received at the hands of Spanish landlords. Many more died during epidemics of European-introduced disease that swept the island during the 1510s and 1520s. In addition, numerous Indians died fighting to protect their lands against European intrusion.

Gold deposits—eagerly wanted and soon discovered by the Spanish—were found in only a few areas of the island. Early colonists were not willing to settle an area that did not have an economic future bright with the promise of mining precious metals. Soon, however, word reached the Spanish colonists of Puerto Rico that rich gold deposits had been found in South America. Upon hearing this news, many of them fled the island and moved to Peru and other Andean countries. Once the island's minerals played out, colonial authorities turned their attention toward growing crops for export. During the sixteenth century, Europeans living in Puerto Rico and surrounding territories realized that sugar cane was a lucrative cash crop. Sugar enjoyed great demand in Europe. Raising the crop required great amounts of hard labor in the sweltering tropical sun, however. Thus began the era of extensive trade in African slaves.

As wealth from the Spanish New World began to flow back to the European homeland, the strategic importance of Puerto Rico was soon recognized. Naval and other military people working in the Americas recognized that the island was a strategic gateway to and from the riches of South America. Puerto Rico grew in importance as a provisioning port for sailing vessels crossing the Atlantic. Ships often brought soldiers and their equipment to the New World and returned with treasures from Latin American mines and sugar fields. In addition, the island was perceived by the Spanish to be a barricade. Because of its location, Puerto Rico was in a position to stop other competing European nations from acquiring the riches of Central and South America. Privateers, known as pirates or buccaneers, came by ship from various European nations to intercept and attack Spanish ships carrying treasures back to the motherland. To protect Spanish Caribbean interests from these privateers, Spain started the construction of impressive military facilities on the island of Puerto Rico.

In actuality, many of the lawless privateers were sponsored by European governments interested in weakening the Spanish economy. Spain soon realized that Puerto Rico must be defended from these bold and frequent attacks. To defend the island, Spanish authorities ordered the island's male residents between ages 16 and 60 to join paramilitary groups. These militias were required to assist Spanish soldiers in defending the island's coastal cities. From the sixteenth to eighteenth centuries, construction of fortresses and defensive walls consumed much of Puerto Rico's limited human and material resources. At the completion of these military projects, thick walls with only a few gates surrounded the capital city of San Juan. These gates were closed at night. On several occasions, the Dutch, English, and French tried to invade the island. Their attacks failed, however, because of the fierce determination of island residents to protect themselves. From behind San Juan's imposing fortifications, the Spanish were able to repel such attacks.

SOCIAL CLASSES DURING THE COLONIAL ERA

During the second half of the seventeenth century, Puerto Rico's sugar industry was unable to compete with producers on other Caribbean islands. For this reason, the island's rural

Founded in 1521, much of Old San Juan is fortified by walls that served to protect the city from foreign intruders between the sixteenth and eighteenth centuries. Shown here is Castillo de San Felipe del Morro, one of several forts that was constructed to help defend San Juan Bay.

geography began to change. Settlement patterns and landscapes once dominated by sugar plantations began to diversify. Livestock ranching, for example, began gaining in importance, as cattle grazed on lush tropical grasslands. During that era, the owners of plantations and livestock ranches were the privileged members of Puerto Rico's society. The most important positions in government, army, and church, though, were held by persons born in Spain. White persons born on the island—known as *Creoles*—were not allowed to occupy important positions, regardless of their economic prosperity or accomplishments.

Spanish-born colonists were at the top of the island's social class pyramid. The second-highest position was reserved for wealthy local merchants and landlords. Occupying the third position on the social pyramid were individuals of mixed ancestry (in many cases the children of white and black parents). This group of multiracial origin usually practiced such occupations such as carpentry, masonry, metallurgy, painting, food processing, shoe making, and textile production. Women of mixed background worked as housemaids, cooks, or dressmakers, or took menial jobs such as washing clothes and ironing. Wealthy families lived inside the capital in houses made of brick and stone, whereas the poor lived in shacks located in shantytowns surrounding the main city.

During the eighteenth century, outside the capital city, cattle ranchers were the most powerful group. Conflict between ranchers and poor peasant farmers was common because farmland was frequently invaded by livestock. Ranch owners did not want to see the development of other agricultural activities. They were fearful of losing their privileged status in the island's countryside. During the nineteenth century, however, the owners of sugar cane plantations regained their powerful former social and economic status.

Many rural residents did not own the land upon which they lived. Being landless, they were forced to enter into agreements with powerful landlords in order to find a place to live. Some of the agreements required the landless farmers to share their crops with the landowners. In other cases, the agreement required the landless person to work for the landlord. Most of the people living in the interior were landless and constantly on the move in search of new places to settle and farm. They had few possessions, permitting them to move easily. Their basic belongings were hammocks for sleeping and pots for cooking. For utensils, they used dried shells of coconuts and pumpkins. The landless class consisted of Creoles with no immediate connection to Spain, blacks, and persons of mixed ancestry.

African slaves occupied the lowest class in Puerto Rico's social pyramid during the colonial era. The number of slaves living on the island was smaller than the number of free blacks, colonists of mixed racial origin, and whites, however. In 1787 the island's population was 103,051, of which slaves numbered 11,260, or only 11 percent of the total. Six decades later, in 1846, the slave population still represented only 11.6 percent of the island's population. Compared to population statistics of other islands, Puerto Rico's slave population was relatively small. Slaves worked mainly on the island's sugar plantations. Living conditions were miserable there, and they received minimum health care and protection. It is little wonder that between 1795 and 1848 these African workers tried 22 times to gain their freedom by engaging in uprisings. Slavery in Puerto Rico was finally abolished in 1873, after long years of struggle by the abolitionist movement supported by liberal politicians in Spain.

SPANISH CONTROL OF THE TERRITORY

Spain maintained tight control of Puerto Rico's economy. The colony was allowed to maintain only one port, which was located in the capital city of San Juan. Puerto Rico also was barred from trading with non-Spanish territories. For these reasons, the island and its residents faced serious economic problems during the colonial era. San Juan received only a few merchant ships annually because of the trade limitations imposed by the Spanish Crown. Consequently, the planters and merchants increasingly engaged in illicit trade with other countries in order to get around Spain's trade monopoly. Clandestine trade took place throughout the island, with even some corrupt Spanish officials participating.

The Spanish authorities faced serious difficulties in attempting to stop illegal businesses. Residents of the colony were in desperate need of products that Spain refused to provide. For this reason, many islanders again turned to illicit trade. Illegal businesses allowed colonists to sell their agricultural products and leather from the cattle ranches to neighboring non-Spanish territories. In addition, residents were able to buy textiles, clothing, oil, flour, wines, and manufactured goods from non-Spanish areas. Puerto Rican contrabandists or smugglers sold horses, cattle, mules, and exotic woods to neighboring islands that lacked these products. This illegal trade may seem immoral, but this form of exchange helped Puerto Ricans survive economically during a colonial era that was characterized by restrictions. All aspects of colonial administration, it must be remembered, were in the hands of Spanish military personnel who had little concern for the well-being of island residents.

During the first three decades of the nineteenth century, Spain lost most of its colonial possessions in the Americas. For this reason, the Spanish government changed its approach toward Puerto Rico. It adopted a more flexible position in order to gain the islanders' sympathy. Essentially, Spain was fearful of losing more territories. In order to avoid revolutionary movements in Puerto Rico, the Spanish authorities invited islanders to vent their frustrations as formal protests to the government. During the second decade of the nineteenth century, an islander by the name of Ramon Power Giralt served as the colony's emissary to the Spanish *Cortes* (National Assembly). He expressed the discontent Puerto Ricans felt toward the militaristic and monopolistic administration responsible for the island's isolation and widespread poverty. The chief complaints that Giralt brought forward dealt with the social and economic conditions of his homeland. Puerto Rican politicians constantly criticized the Spanish taxation system and trade restrictions. They also were critical of limitations on Puerto Rican participation in the administration of the island, the lack of schools, poor health care, and the inefficiency of public infrastructure.

INDEPENDENCE MOVEMENTS

At first, the Spanish government was willing to accept some of the demands made by Giralt and other Puerto Rican liberals.

The liberals worked to improve the social and economic conditions of the island and toward gaining civil and political rights. As the nineteenth century progressed, however, conditions in Spain became increasingly unstable. This resulted in little being done to improve matters in Puerto Rico. In fact, Spain appointed governors who were oppressive. Puerto Ricans, frustrated with their political situation, began to organize separatist groups.

From 1809 to 1868, the Spanish authorities discovered eight conspiracies organized to start uprisings that would hopefully lead to the island's independence. Many Puerto Rican pro-independence leaders were forced to leave the island. Others sought refuge in the island's rugged interior. Soon, separatist organizations sprouted up in villages and towns across the colony. On September 23, 1868, several hundred men took a town in the central highlands of Puerto Rico. They arrested its officials and proclaimed the independent Republic of Puerto Rico. However, the new independent country would be short lived. As soon as the Spanish authorities got word of what had happened, they sent military forces to the island's interior and won an easy victory over the revolutionaries. Hundreds of Puerto Ricans were tortured because of their sympathies toward independence.

Even though the rebels were viewed as anarchists by the islands' elite, who were generally loyal to the Spanish Crown, these revolutionaries had noble intentions. Their first goal, for example, was to do away with slavery. Slavery was eventually abolished, in 1873. The 1868 revolution was brief but significant: It marked the beginning of the idea that Puerto Ricans were members of a separate nation, even though the island had never been independent. Today, Puerto Rico is an American territory, and its population enjoys American citizenship. But when an islander is asked to identify his or her nationality, the answer invariably will be "Puerto Rican."

In 1898, 30 years after the pro-independence insurrection, the Spanish government granted autonomy to Puerto Rico.

Historical Geography

In July 1898, during the Spanish-American War, U.S. armed forces landed more than 3,000 troops at Guanica and quickly captured Puerto Rico. Shown here is Battery B of the Fourth Artillery during an attack on Coamo in August 1898.

This was the first system of home rule the island had ever known. Under the autonomy agreement, Puerto Rico achieved almost all the privileges of an independent country. As it turned out, Spain's granting autonomy to Puerto Rico came too late for the island to hold onto it. The country enjoyed only a few months of self-rule. In July 1898, U.S. Naval forces occupied the territory. Puerto Rico became an American territory once Spain was defeated by the United States during the Spanish-American War. This war was a brief conflict between Spain and the United States arising from Spain's abusive policies in Cuba. It was, to a large degree, brought about by the efforts of U.S. expansionists. Essentially, the United States was interested in having a strong presence in the Caribbean. Spain was seen as a corrupt empire that was unworthy of having colonies in the Americas.

PUERTO RICO UNDER THE AMERICAN FLAG

Once American troops landed on the island, the Puerto Ricans were divided about how to receive their new conquerors. Some welcomed the American occupation, but others resented becoming a possession of the United States. Some opponents of America, however, began to see advantages in the island's new status. They thought the United States would actively improve communications, transportation, education, and health services on the island. In 1917, after 18 years of U.S. occupation, Puerto Ricans gained American citizenship and a greater degree of self-government. The United States appointed the island's governors, however. This changed in 1952 when the United States allowed Puerto Ricans to have local home rule through popular election of a governor and members of a Senate and a House of Representatives. The territory became known as the Commonwealth of Puerto Rico. Its citizens may serve in the American military but are not obligated to pay income tax. They are not allowed to vote in national elections and do not have voting membership in the United States Congress.

During the first decades of the twentieth century, the majority of Puerto Rico's labor force was engaged in agriculture. Salaries were very low. These workers had serious difficulties meeting their daily needs, and their housing conditions were deplorable. It was difficult for them to buy medicines and decent food. Diets were poor. Rice and beans was the regular meal, and on special occasions codfish was included. Agricultural workers toiled for twelve hours a day in the sugar and coffee plantations, regardless of intense heat, heavy rain, or insect plagues. For these reasons, the rural population suffered numerous health problems.

Boys and girls were forced to work at an early age, helping their parents with the domestic chores and the agricultural work. For example, boys worked in small home fields that produced crops for the family's consumption. Rural boys also had to fetch wood and water and take care of the family's livestock, if, indeed, the family had money to purchase animals. Girls were in charge of house cleaning, cooking for the family, and taking care of their younger siblings. In addition, these boys and girls worked in the fields collecting coffee and providing water to the workers. Their houses were often made of cardboard, poor quality wood, palm tree leaves, and rusted panels of light steel. Entire communities of these fragile shacks were often destroyed by hurricanes, which are common to the region.

Sugar plantation workers lived in one-bedroom small houses with no space for planting a garden to feed their families. These houses did not have a bathroom. Families had to share public restrooms provided by the sugar companies. Kitchens were located outside the house, and there was no water and electricity in the homes. Many of these homes were close to the sugar processing facilities. Residents were exposed to the fumes and loud noises of the sugar mills. Life in the Puerto Rican countryside was difficult. Many families tried to escape the misery of the countryside by moving to the city in search of a better life. Nevertheless, in the majority of cases, they ended up living in overcrowded shantytowns built around the cities. Many Puerto Ricans who could afford passage migrated to the mainland, particularly to New York City. As United States citizens, they were granted automatic entry.

MODERNIZATION OF THE ISLAND

Something had to be done to fight the poverty that prevailed in Puerto Rico before the 1950s. In response to this need, the island's public administrators developed a program of industrialization that would lead to the modernization. During the 1940s, U.S. and Puerto Rican authorities became active sponsors of various industrial enterprises. The government helped in construction of factories and encouraged investment in industrial ventures. New roads, bridges, power plants, and planning agencies were developed to support rapid industrialization. This modernization program was called "Operation Bootstrap."

"Operation Bootstrap" began with the government's construction of a modern cement plant. This facility was essential for construction of the factories, roads, and low-cost housing that were part of the modernization program. Part of the plan was to attract foreign investment. To achieve this goal, companies were offered building space and a ten-year exemption from the payment of taxes. These incentives were successful in luring many American companies to the island. By 1956, industry and services replaced agriculture as the island's leading source of revenue. Industry was diverse, making it less vulnerable to economic decline in one particular sector. The island produced electronics, rubber products, building materials, synthetic textiles, frozen foods, and clothing. As the island's population became better trained, numerous pharmaceutical companies established operations there. Today, however, Puerto Rico is beginning to lose many of these industries. Corporations are turning to Southeast Asia and elsewhere in Latin America, where labor and other operation costs are less expensive.

"Operation Bootstrap" also helped increase tourism. The government built a number of modern hotels on the island. With the development of modern facilities, Puerto Rico rapidly became the most important tourist destination in the Caribbean. Now, however, many other Caribbean islands have developed resort facilities of equal or superior quality. Puerto Rico no longer has the tourist destination appeal that it held from the 1960s through the 1980s. Travelers can find better "deals" outside Puerto Rico, because hotel operation costs are lower elsewhere in the Caribbean. Once again, the government is stepping in, hoping to boost this sector of the economy. For example, it is offering handsome incentives to cruise companies if they dock in Puerto Rico. In addition, Puerto Ricans are trying to preserve the island's few pristine forests and coastal areas. They hope that by doing so, people interested in ecotourism and the preservation of natural habitats will be attracted to the island.

POLITICS DURING THE LAST HALF OF THE TWENTIETH CENTURY

Even though Puerto Ricans cannot vote in American presidential elections, many of them are quite content with their semi-autonomous commonwealth political status. A chief reason is that as long they live on the island they are exempt from all federal income taxes. In addition, islanders benefit from roughly $10 billion in annual federal spending.

During the 1990s, Puerto Ricans twice had the opportunity to vote on the issue of statehood, independence, or retaining commonwealth status. On both occasions, the commonwealth option prevailed. Many of those voting to reject statehood did so on cultural grounds. They were afraid that with statehood they would lose Spanish as the island's official language. They wanted to retain the island's distinctive Puerto Rican and Hispanic cultures. The first two gubernatorial elections of the twenty-first century were won by the pro-commonwealth, Popular Democratic Party. This is the party that was founded during the late 1930s by Luís Munoz Marín, the driving force behind "Operation Bootstrap."

CHAPTER 4

People and Culture

Puerto Rico's highways are some of the best in Latin America, but they are also some of the busiest. During the workweek, all roads leading to or from the cities are crawling with bumper-to-bumper traffic. The crowded highways underscore the density of Puerto Rico's population. The most heavily populated municipality on the island is its capital city. San Juan has a population density of about 9,300 people per square mile (5,770 per square kilometer). Only in remote towns between the lush hills of central Puerto Rico are population densities as low as 400 persons per square mile (250 per square kilometer).

Factories, residential areas, and modern hotels now occupy the coastal lowlands, where sugar cane used to be the king of the island's agriculture and plantations dotted the fertile landscape. During the 1950s and 1960s, modern factories invaded the lowland countryside. Lured by factory jobs, poor farmers fled the interior highlands in

Today, there are more than one million motor vehicles that travel on Puerto Rico's roads and highways. Shown here is a busy street in Old San Juan, Puerto Rico.

search of a better future on the coast. Housing had to be built to accommodate the new industrial workers. Large residential complexes were developed in coastal areas close to the capital and manufacturing facilities. In the process, much of the northern coast was transformed from a tropical paradise into a concrete jungle. During that same era, many communities in the verdant hill country experienced declining populations.

The development of better highways, coupled with the newfound prosperity industrial workers enjoyed, motivated many Puerto Ricans to buy automobiles. Today, the small island has more than one million motor vehicles. It is little wonder that the island has the busiest roads in the Caribbean.

To further complicate matters, 94 percent of the island's people live in cities, most of them in and around San Juan.

High population densities and crowded spaces have not always characterized Puerto Rico's population geography. During the early days of Spanish occupation, it was very difficult to convince the Spanish to stay on the island. Constant Amerindian uprisings and attacks by European pirates discouraged early colonists. Consequently, when many of these settlers heard the news about mineral riches of Mexico and Peru, they quickly abandoned the island. Only a few white colonists with their African slaves, and some Amerindians who managed to survive their conquest, stayed in Puerto Rico. For three centuries, the island remained a poor colony. Spain's only interest in the island was strategic. It valued the huge fortress built on the island's north shore, needed to protect the Hispanic Caribbean from intrusion by other European powers.

Because the Spanish controlled the island but were not interested in settling on it, population growth was very slow between the 1530s and 1790s. This situation changed during the nineteenth century. During the 1800s, Spain decided to focus its attention on the only two territories that they retained after the wars of Hispanic American independence. In order to maintain possession and improve the economies of the colonies of Cuba and Puerto Rico, Spain passed several laws. The Spanish and other Europeans from Roman Catholic countries were invited to live in the Caribbean. In addition, Spanish authorities were willing to grant freedom to blacks from non-Hispanic colonies as long they demonstrated loyalty toward Spain. Many runaway slaves from other islands in the Caribbean chose Puerto Rico as their new home.

Other factors contributed to the increase of Puerto Rico's population during the nineteenth century. Sanitation practices were adopted, important advances were made in tropical medicine, and vaccines were introduced. For these reasons, death rates dropped while the population increased. Because birth rates were high from 1899 to 1950, the island experienced a "population explosion" that transformed Puerto Rico into one of the most crowded places in the Western Hemisphere.

Puerto Rico's population of 3,900,000 people is ethnically and racially diverse. Most are from Spanish or African descent. A small number of individuals of remote native Indian ancestry continue to live on the island. Since the sixteenth century, the island has welcomed various groups. Immigrants have come from elsewhere in Latin American and from many European countries. Some Middle Easterners also have blended in with the local population. Race relations on the island are cordial, and intermarriage among different racial groups is common. For example, in Puerto Rico, whites and blacks intermix much more than they do in rest of the Americas, with the possible exception of Brazil.

NATIVE GROUPS

Puerto Rico's earliest inhabitants were Amerindian peoples who arrived in waves of migration that began perhaps in 2000 B.C. Some archaeologists believe that the island's first natives came from North America. These earliest arrivals were eventually displaced by Amerindian groups from what is now eastern Venezuela. The group that came from Venezuela was known as the Igneri. They were later displaced by the Taino, whose culture was part of the Arawak cultural complex of northern South America. When the Spaniards arrived in the region, the Taino people were the dominant native group in the western part of the Caribbean. The Taino led a sedentary lifestyle, living on the coast and along the rivers in villages of huts known as *bohios*. They farmed the land, made primitive stone tools and weapons, and baked clay pots. Archaeologists and anthropologists estimate that about 60,000 Tainos were in Puerto Rico when the Spanish explored the area for the first time. By the 1530s, only 1,553 remained; the rest had succumbed to European-introduced diseases or to brutal treatment. As the native population began to dwindle, African slaves were brought to the island to work the plantations. Eventually, most Tainos married Spanish settlers, and their culture finally died out.

Although Puerto Rico's exclusively native population disappeared after the sixteenth century, many Taino cultural traits were adopted by the Spanish. Today, one can find numerous words of Amerindian origin in use in the Puerto Rican culture. In addition, many places in the island have Indian names. Taino agricultural products such as manioc (also known as cassava) and guava fruits were embraced by the Spanish settlers. Puerto Rican popular music uses percussion instruments of Taino origin; these include the maracas and the guiro. The latter is the dried shell of a fruit similar in shape to a cucumber. It is scratched with a stick or narrow metal rod in order to produce a melodic sound. The Spanish also learned about hammocks from the Taino. The most interesting remains of Taino cultural landscape are the *bateyes,* or ceremonial ball courts, and today two of them are popular tourist attractions. Religious celebrations and a ball game, which resembled a rough version of soccer, took place in these sites during the Taino era.

THE AFRICANS

As the Taino died out, new hands were needed to work the land. Beginning in the 1520s, slave traders transported African slaves to the island. So many arrived that by the 1530s, Africans had become the island's dominant ethnic group. The majority of the slaves came from the region between the Niger and Congo rivers in West Africa. In 1802, persons of African descent made up 52 percent of Puerto Rico's population.

Today, most of the island's blacks live on the northeastern coast. The town of Loiza is famous for the preservation of African traditions. In this town, evidence of the island's slave-trading days is impossible to ignore; the majority of the residents of this community are black. Some African rituals are still practiced here. In fact, Loiza is believed to be one of the purest

People and Culture   51

Each summer, residents of Loiza, a black community on the northeastern coast of Puerto Rico, celebrate their African heritage during the Santiago Apostol Festival. The lively celebration includes costumes, masks, and bomba dancing.

centers of true African traditions in the Americas. During the summer, this black community celebrates its heritage with a lively festival. Residents dress in ceremonial costumes similar to those of the Yoruba communities of West Africa, with whom many of the island's blacks share ancestry. Another part of the island that has a significant black population is the southeastern lowlands. Essentially, any of the coastal communities where sugar was harvested during the eighteenth and nineteenth centuries is likely to have a significant black population.

EUROPEANS

The island's 3.9 million inhabitants are mainly Hispanic in culture, with varying racial backgrounds. All are Spanish-speaking, and two-thirds are nominally Roman Catholic. Many who think of themselves as "white" point to their descent from the Spanish who came to the island during the sixteenth century, or came later to establish plantations. An elite group of white landowners owned much of the island's best land. The rest of Puerto Rico's Spanish population became tenant farmers on small land holdings. With the establishment of sugar cane plantations, the white population of low socioeconomic status was forced out of the sugar districts. White workers were an expensive source of labor; they could not compete with African slaves, who received no wages. Without land or jobs, the poor whites fled to the central highlands to settle as unlawful tenants on the unfenced cattle ranges of large landowners. Subsistence farming was their main occupation.

During the first half of the nineteenth century, Puerto Rico received a significant number of white immigrants. Many came from the poorer districts of Spain or elsewhere in Mediterranean Europe. These immigrants selected the island as their new home for several reasons. Of great importance was the fact that this territory was one of the few Spanish possessions in the Americas in which revolutions, dictatorships, and political instability had not developed. Some of these new colonists came from former Spanish possessions elsewhere in Latin America. Many had been expelled from their previous home because of their loyalty to the Spanish government.

THE BLENDING OF THREE RACES

Throughout Puerto Rico's history, interracial (and intercultural) mixing has been commonplace. During the colonial era, it was not uncommon for a Spanish man to live with an Amerindian or African woman. In addition, unions between slaves and Indians were common. The term *mestizo* is used to describe offspring of mixed Spanish and Native American ancestry. First-generation persons born from the union of white and black persons are called mulattoes. The union between mulattoes and mestizos was so common that eventually most of the island's people did not fit in the white, black, or Amerindian racial categories. This mixing was not limited to racial characteristics. It also included an exchange and blending of culture traits, such as language, diet, art, values, and beliefs. The black, mulatto, and mestizo populations, separately or in combination, still make important contributions to the Puerto Rican culture. Puerto Rican folk music, for example, combines Spanish, African, and Amerindian rhythms. The popular music of the island also has benefited from this fusion. Some of the best groups playing tropical salsa music are from Puerto Rico. Devoted admirers of these musicians can be found throughout the Caribbean and elsewhere in Latin America, but also in Europe and the United States.

Some religious traditions in Puerto Rico demonstrate the fusion of the island's cultures. Many Puerto Ricans practice Santeria, which is a form of African folk Catholicism wherein Yoruba deities are identified with Catholic saints. In theory, followers of Santeria are supposed to adore images of Catholic saints. In practice, however, the images are really representations of African gods called Orishas. The practitioners pray to the Orishas in exchange for favors and blessings. On the island, stores known as *botanicas* sell artifacts used in Santeria rituals. These stores demonstrate the powerful influence of African religions in Puerto Rico's popular culture.

LANGUAGE

Most Puerto Ricans speak Spanish, the language passed down from the days of Spanish colonial rule. The Spanish spoken in Puerto Rico is well laced with Amerindian and African words. One of the characteristics of Puerto Rican Spanish is that in words ending in "s," the "s" sound is muted and sometimes disappears altogether. In addition, many Puerto Ricans exchange "l" for "r," and vice versa. A small number of residents speak English. Most of these are Americans holding a government position, such as the military or some other institution. Others are engaged in the business activities of multinational corporations. Many of the better educated Puerto Ricans, as well as those involved in tourism and other service industries that cater to outsiders, are multilingual.

After the United States occupied the island in 1898, English began creeping into Puerto Rican Spanish. A mixture of the two languages is common, in a version popularly called *Spanglish*. Initially, this mixing may have been influenced when labels on consumer products were written in English. For example, the Spanish word for diapers is *pañales*, but in Puerto Rico many persons use the word *Pampers* to denote this product for babies. Another example is the use of the brand name *Clorox* instead of the Spanish term for bleaching agent, *blanqueador*. The correct Spanish term for corn flakes is *hojuelas de maiz*, but islanders prefer to use *corn flakes*. A hot dog in Spanish is a *perro caliente*, but Puerto Ricans say "hot dog." Puerto Rico's Spanglish vocabulary developed momentum after thousands of Puerto Ricans living in the United States returned to the island from the 1960s to the 1980s. These returnees imported many English words and speech patterns from New York and elsewhere in the northeastern United States.

RELIGION

Under Spanish colonial occupation, the Roman Catholic Church played a key role in the cultural, social, and political geography of the island. For example, most of the island's schools and hospitals had some link with the Catholic Church. With the establishment of an American administration on the territory, Catholic institutions lost the preferential treatment they enjoyed under the Spanish Crown. As the American presence was established, numerous Protestant churches were opened on the island. However, the island's first Protestant church dates to 1873 and was built in the southern city of Ponce. This place of worship was an Anglican Church. The group arrived soon after the Spanish Crown decided to allow religious groups other than the Catholic Church to operate on the island. By 2001, about one-third of the island's population considered themselves Protestants. Puerto Rico reflects a growing trend throughout Latin America toward religious diversity.

5

Political Geography

During the last decade of the nineteenth century, the United States government demonstrated significant interest in the Caribbean and in Central America—together, with Mexico, known as Middle America. This interest was based on the region's strategic position. Middle America was the bridge linking two continents, North and South America. It also held the narrowest barrier separating the Pacific and Atlantic Oceans. Further, American corporations began to establish plantations in Middle America, and the region became a significant consumer of United States products. American politicians believed that it was important for the United States to have control over parts of Middle America because the region was of growing economic and strategic political importance.

By the 1890s, United States entrepreneurs had invested millions of dollars in Cuba and Puerto Rico. The United States decided to take advantage of the political instabilities that Spanish Caribbean territories were experiencing during the late 1800s. Essentially, Americans wanted to profit from Spain's difficulties. In order to declare war on Spain, the United States had to have an "excuse." It used as a pretext the mysterious destruction by explosion of the American ship *Maine* in Havana, Cuba, on February 15, 1898.

American conspiracy theorists attributed the destruction of the *Maine* to a Spanish bomb. Investigations, on the other hand, indicated that the cause of the tragedy might have been a faulty boiler. Spain did not want to go to war, declared its innocence, and offered reparations for the *Maine* incident. Nevertheless, the United States wanted a war that would let it gain control over the strategic Caribbean region. The war with Spain lasted less than four months. At the close of the conflict, Puerto Rico was ceded to the United States. Puerto Rico's local politicians were not consulted, and this was the first time the United States extended its territory beyond the North American continent. Many islanders saw the shift to American rule as a sad setback to their long quest for Puerto Rican self-government.

U.S. OCCUPATION

The defeated Spanish left Puerto Rico in bad condition. American occupation forces immediately began a series of port developments, road constructions, and agricultural projects. They hoped to attract and benefit U.S. companies interested in establishing businesses in the territory. During the first four decades of the twentieth century, the United States allowed Puerto Rico only very limited self-government. A Puerto Rican House of Delegates was formed and elected, but the island was governed directly from Washington. United States presidents appointed governors from the mainland, and most had little knowledge of the culture and traditions of Puerto Rico. Some of these American administrators treated the islanders condescendingly. The position of the U.S. government from 1898 through the 1940s was that Puerto Ricans had no right to self-government. During that period, the U.S. Supreme Court was the ultimate arbiter of the Puerto Rican legal system. The main opinion of American politicians from that era was that Puerto Rico was United States property. Essentially, they saw Puerto Ricans as conquered people.

For the first 15 years of American control over the island, many Puerto Rican politicians complained bitterly. They were outraged over what they believed to be a lack of essential democratic rights. During that period, the U.S. government was not too concerned about the rights and well-being of Puerto Ricans. It was quite anxious, however, about the growing political instability in Latin America and Europe. For this reason, American priorities were linked to anything that could reinforce U.S. military and commercial presence on both sides of the Atlantic. Military analysts considered the Caribbean as critical to U.S. defense. Some European powers hostile to North America were promising assistance to Latin America in exchange for military or political alliance.

Because of its strategic location, Puerto Rico and its neighboring islands were thought of as the key to the Caribbean. During the 1910s, the United States purchased the Virgin Islands, located just east of Puerto Rico, from Denmark. With the purchase, the United States hoped to forestall any possibility of a German presence in the area. A few weeks before the United States declared war on Germany, Congress passed the Jones Act. This legislation granted (imposed) American citizenship on all Puerto Ricans, with the exception of individuals with authorization to turn down this status. American politicians of that era believed that citizenship had to be imposed as a way of emphasizing that Puerto Rico was a permanent territory of the United States. These same politicians wanted to eliminate any anti-American movement in the area, and they thought that by forever linking Puerto Rico to the United States, hostile sentiments would be erased.

The Jones Act had some contradictions. Puerto Ricans were subject to the military draft, but they could not be represented in Congress or vote in U.S. presidential elections. In addition, islanders could not be elected as governors of Puerto Rico. A local Senate and House of Delegates were established, but its legislation could be rejected by the American governor. Particularly odious to many Puerto Ricans was the creation by the Jones Act of an administrative position called The Auditor. This official, appointed by Washington without Puerto Rican consent and given exclusive and unilateral powers, had the right to examine all accounts of the island's government.

Despite the political limitations linked to the Jones Act, a significant portion of the island's population was optimistic that a positive association with the United States would eventually develop. They never gave up hope in this regard. A few Puerto Ricans started agitating for independence, violently challenging the United States presence. From the 1930s to the 1950s, some of these pro-independence agitators were jailed. During this time, however, a new generation of politicians arrived on the island's political scene. They were progressive and started a campaign aimed at developing a politically fair and economically healthy association with the United States. Among these progressive leaders was Luís Munoz Marín, who founded the Popular Democratic Party during the late 1930s. Marín was down-to-earth; he believed that the feud between pro-independence and pro-statehood Puerto Ricans was a waste of energy. He believed that it was more important for islanders to work together to bring economic development to the island.

One of the most important aspects of Puerto Rico's 1932 local elections was the participation of female voters. During the first decades of the twentieth century, only literate male citizens were allowed to vote in local elections. However, since 1917, the Puerto Rican Suffrage League had been campaigning in favor of women's participation in political activities. This league consisted of female islanders who had studied in the United States and had contacts with American feminist organizations. The league faced many opponents. Particular opposition came from

Luís Munoz Marín, who founded the Popular Democratic Party during the late 1930s, served as governor of Puerto Rico for four terms (from 1948 to 1964). Marín, shown here in 1957, supported his country's economic development and helped Puerto Rico achieve commonwealth status.

male politicians, intellectuals, and religious leaders who thought that a woman's place was in the home. Members of the league, however, stressed the fact that the island's women were better prepared to understand the problems related to education, motherhood, and child development than were men. Unfortunately, the league was somewhat exclusive in that it believed illiterate women should not be allowed to vote.

Blue-collar women who were not part of the aristocratic Suffrage League made important contributions to the women's voting rights movement. During the early 1900s, they demanded participation in local elections. They hoped that by being politically active they could improve their working conditions and fight exploitation and poverty. Agitation for women's voting rights was modeled on similar movements that had been successful in other parts of the world. In 1924, the various feminist organizations, including the Proletarian Women's Movement formed a coalition. This alliance proved quite positive, because it gave working-class women the opportunity to join forces with more elitist feminist groups that were fighting for a common cause. By 1932, they had succeeded. Women were allowed to vote, and for the first time a woman was elected to the legislature. Illiterate women, however, were not allowed to vote until 1935.

During World War II, Puerto Rico regained its strategic geopolitical importance. The island became the center of a chain of military bases the United States established from Trinidad, in the Lesser Antilles, to Cuba. Some military analysts called Puerto Rico the "Gibraltar of the Caribbean." This reference was an allusion to the strategic location of the British colony of Gibraltar on the strait that links the North Atlantic Ocean and the Mediterranean Sea. During the early 1940s, many Puerto Ricans lost their land. The American military expropriated sections of the island for the establishment of military facilities built to defend the Caribbean Sea and the Middle Atlantic from possible German attacks. Puerto Ricans fought alongside other American citizens in World War II, in the effort to defeat the Germans and Japanese.

Following World War II, American had almost no interest in granting statehood to Puerto Rico, despite Puerto Rican contributions to the war effort. The United States granted independence to the Philippines in 1946, but most American legislators did not favor an independent Puerto Rico. Perhaps it was the looming shadow of the Cold War that put a damper on the idea of total freedom for Puerto Rico. Numerous political analysts have accused the Puerto Rican people of being insecure or uncertain concerning possible statehood or possible independence. The United States Congress, however, has provided little guidance in the matter, preferring instead to deal with the present reality. That reality is the commonwealth.

THE BIRTH OF THE COMMONWEALTH

The United States government was not sure what to do with Puerto Rico following World War II. Munoz Marín suggested a new political status known as the free associated state, or *Estado Libre Asociado*. The term *free associated state* was quickly changed to "commonwealth." U.S. State Department officials wanted to avoid any confusion caused by the use of the word *state*. The political status Munoz Marín proposed was neither statehood nor independence, but something in between. His political formula proposed an elected governor, rather than one imposed on Puerto Rico from Washington. It delayed plans for independence or statehood, since social reforms and economic growth were more urgent. In 1950, the United States Congress passed legislation creating the Commonwealth of Puerto Rico, and the island's population overwhelmingly endorsed it in a 1952 referendum. The majority of Puerto Ricans found this new status attractive. It allowed them to take advantage of U.S. economic benefits without surrendering their culture and national identity. Under the commonwealth status, the United States government makes all decisions regarding trade, security, and diplomatic policies. Puerto Ricans can carry U.S. passports, and they remain U.S. citizens subject to most federal laws. One disadvantage of this political system is that power to either allow or disallow statehood, or independence, rests with the U.S. Congress. Thus far, however, islanders seem reasonably satisfied with their commonwealth status.

During the 1950s and 1960s, the United States experienced a tense relationship with communist countries. One of the superpower's greatest fears during that period was the emergence of communist and anti-American regimes in Latin America. Social and political injustices, as well as widespread poverty, were common throughout Latin America. In many instances, United States–supported dictators were responsible for the problems affecting the region, and this allowed communism to get a foothold. The United States wanted to change its image from a powerful nation assisting tyrannical rulers to a patriarchal power able to modernize the economies and societies of its allies. The success of its economic policies in Puerto Rico did not go unnoticed in the hemisphere. Nor did it go unnoticed that the United States economy was one of the prime beneficiaries of "Operation Bootstrap."

From the late 1940s to the 1960s, the U.S. government supported many projects aimed at improving the standard of living for Puerto Ricans. The electrical power system was expanded to all municipalities on the island, including the most isolated highland areas. By the 1960s, 90 percent of Puerto Rican homes had electricity. The spread of electrical power motivated islanders to buy electrical appliances such as washing machines, refrigerators, and televisions. The adoption of these appliances made the lives of Puerto Ricans easier. It also transformed the island into an important market for United States-manufactured products. Drinking water distribution systems also were expanded and improved. This helped Puerto Ricans enjoy healthier lives, especially in the countryside where poor sanitary conditions had been the norm.

Housing was another area that improved greatly during the 1960s. By 1940, it was estimated that 80 percent of Puerto Ricans did not have adequate housing. Shantytowns built by people trying to escape the misery that prevailed in the countryside surrounded the island's cities. In 1957, with U.S. assistance, the Commonwealth founded the Housing and Urban Renewal Corporation. This agency built more than 34,000 public housing units during the 1960s. A public housing project apartment had the advantage of low rent fees. In addition to the public housing projects, the government offered low-interest loans and technical advice to those interested in building their own homes on lots provided by the Commonwealth. All these government initiatives were fruitful in terms of improving the structural quality of houses. By 1960, only 20 percent of housing was considered inadequate. The 1990 census revealed that only 12 percent of dwellings did not meet the standards of "satisfactory" housing.

There was a down side to public housing, however. Friends and large families, formerly united in their village or shantytown by emotional bonds, were often separated as workers secured new employment and needed new housing. Moving to housing projects forced them to coexist with strangers and dissolved ties that had formerly bound them to their village or shantytowns. Even though these slums were very poor communities, residents were rural-to-urban migrants who came from the same rural area and sometimes from the same family. In terms of appearance, the shantytown was a miserable place. But its residents were used to living in them and cooperating with friends and relatives. That sense of solidarity and community was shattered in the modern housing projects, leading to social disorientation and often to crime.

EDUCATION

The island's modernization era, from the late 1940s to the 1960s, brought significant improvements in education. During that period, education was the government's top priority. Numerous schools were built, and hundreds of teachers were educated and hired. By 1940, 31 percent of the island's population was illiterate, but by 1960 the rate of illiteracy was reduced to 17 percent. Government reports released in 2002 revealed that 94.1 percent of the population age 15 and over can read and write. These statistics place Puerto Rico ahead of most Latin American countries for literacy. Obviously, the joint effort of local and U.S. governments to educate the island's population have been effective. Today, there are more than 1,500 schools in Puerto Rico, with nearly 42,500 teachers educating more than 650,000 students.

The University of Puerto Rico was founded in 1903. During the 1960s and 1970s, the university underwent many significant improvements including the addition of many new

Founded in 1903, the University of Puerto Rico's main campus at Rio Piedras is located in the southern section of San Juan. Shown here is the university's most renowned symbol, the clock tower, which overlooks the suburb of Rio Piedras.

academic degree programs. Additionally, new branch campuses opened their doors in all of the island's geographic regions. Now, Puerto Ricans interested in earning a college degree no longer need to commute long distances or leave the island. In 2005, the University of Puerto Rico system had 70,000 students and a faculty of 5,000. This public higher education system has 11 academically accredited branches. Three units of this system offer doctorate degrees and are among the most important research centers in Middle America.

The main campus, located in the San Juan suburb of Rio Piedras, includes many colleges and graduate schools, which offer programs in social sciences, business administration, interdisciplinary studies, humanities, law, architecture, education, natural sciences, communications, planning, and information technologies. The University of Puerto Rico's School of Medicine, developed during the 1950s, offers degrees in medicine, pharmacy, dentistry, public health, and nursing. Approximately 3,000 students attend this school. Puerto Ricans interested in earning a degree in engineering from a public institution attend the university's westernmost campus in Mayaguez.

The second most important higher education system of Puerto Rico belongs to a private institution known as *Universidad Interamericana de Puerto Rico* or Puerto Rico's Inter-American University. This university strives to serve as a cultural bridge connecting Northern (Anglo) America with Latin America. Protestant missionaries founded the university system during the 1910s. In 2005, this college system enrolled 43,000 students and employed 2,000 professors. The Inter-American University offers doctorate degrees in optometry, psychology, theology, human resources, education, and law.

HEALTH

The public health program of the 1950s was among the most significant achievements of the United States and Puerto Rican government modernization initiatives. The island's Department of Health and the Planning Board joined forces to provide health services to all geographic regions of Puerto Rico. These organizations made sure that services were accessible to the majority of the island's population by dividing the country into five regions. Each region centered on a regional hospital and a central office in charge of managing disease prevention plans and operating healthcare programs. In addition, government assistance enabled municipalities to operate their own clinics.

The Puerto Rico Department of Health fosters the physical and mental health of the island's population. It attempts to help Puerto Ricans enjoy healthy and productive lives. That this government agency has diligently pursued its goals is reflected in statistical comparisons. In 1964, the infant death rate was 52 deaths per 1,000 live births. Ten years later, this rate dropped to 23 deaths per 1,000 live births. In 2002, it was 9.8 deaths per thousand. By contrast, the Dominican Republic, Puerto Rico's closest Hispanic neighbor, has an infant mortality rate of 47. The life expectancy in Puerto Rico is another index that demonstrates progress on the island. In 1940, life expectancy in Puerto Rico was only 46 years. By 2005, life expectancy on the island had risen to 77.5 years, exceeding the 77-year average in the United States!

In 1940, the island had only 456 physicians. By 1965, that number had increased to 2,267, and by the year 2000 to 7,800. The increase in the number of medical doctors during the 1950s was the result of a U.S.-government-sponsored program. Scholarships were granted to licensed general medicine doctors who wanted to become specialists in some branch of medicine. U.S. funds also helped develop programs designed to control tuberculosis, cancer, and heart disease.

The island's most important healthcare facility is the Puerto Rico Medical Center, located in the capital's suburb of Rio Piedras. This facility was opened to the general public in 1964 with the goal of offering specialized medical services not available elsewhere on the island. For example, the island's first oncology clinic, which treats cancer patients, was part of this medical complex. In addition, the Medical Center incorporated within its campus a specialized clinic for the treatment of severely burned patients. The quality of medical treatment improved significantly when the University of Puerto Rico built its School of Medicine near the island's Medical Center. The Cardiovascular Center, which deals with heart and blood vessel problems, was inaugurated during the early 1990s. This institution is affiliated with the University of Puerto Rico Medical Sciences Campus and offers quality cardiovascular care for residents of the island and of neighboring countries.

THE PRO-STATEHOOD MOVEMENT

During the 1960s, the island's pro-statehood movement gained strength, and the New Progressive Party was founded. This political body was formed by local politicians who recognized that the commonwealth status was politically limiting. It did not allow Puerto Ricans to vote for the U.S. president or to have senators or representatives in the U.S. Congress. The main goal of the New Progressive Party is to make Puerto Rico the fifty-first state of the American Union. In November 1993, Puerto Ricans took a referendum, without the total approval of the U.S. Congress. The statehood option was favored by 46.2 percent of the voters. In a similar referendum in 1967, only 38.9 percent of the voters favored statehood. By comparing these figures and noticing the increasing trend, it is easy to predict that, with time, the pro-statehood option will gain more support.

Those who embrace the pro-statehood movement believe that a permanent union with the United States would resolve the inequities that stem from Puerto Rico's Commonwealth status. Puerto Rico would be represented in Congress and the Senate, and Puerto Ricans could vote in U.S. presidential elections. If the island were to become a state, though, it would be entitled to seven or eight representatives in Congress. This would make Puerto Rico politically more influential than about 20 other states. For this reason, some politicians from sparsely populated U.S. states have opposed the Puerto Rican statehood movement.

Political Geography

During the 1960s, a pro-statehood movement gained strength in Puerto Rico and since that time there have been three referendums (1967, 1993, and 1998) to explore how many citizens favor joining the Union as the 51st state. Shown here is a young woman in 1998 who is displaying a banner in support of statehood.

Language poses another obstacle to statehood. The vast majority of Puerto Ricans use Spanish as their primary language, and few islanders speak English fluently. Many American politicians believe that if Puerto Rico wants to be part of the Union, the islanders must use English in its schools. Those opposing statehood are afraid that if English became the island's main language, Puerto Rican culture would be eroded.

In many ways, Puerto Rico is a showcase of democracy. More than 70 percent of those eligible to vote participate in elections and referendums, a figure much higher than in the United States. It remains to be seen, however, whether Puerto will become the fifty-first state anytime in the near future.

POLITICAL ORGANIZATION

Puerto Rico's government functions under its own charter, adopted in 1952. At the head of the island's government is an elected governor. The governor serves a four-year term and can be reelected. The governor's cabinet is a council of executive secretaries (directors) who are appointed by the governor with the consent of the island's senate. Secretaries are the managers of departments such as the state, justice, education, health, internal revenue services, labor, commerce, public works, and others.

The Commonwealth Senate consists of 27 members (two for each of the eight senatorial districts and eleven from the entire territory, rather than from districts). Like the governor and members of the House of Representatives, the senators are elected every four years. The island's House of Representatives consists of 51 delegates, one for each district (the Commonwealth has 40 representative districts) and 11 from the entire territory. Two political parties have dominated the government. According to law, however, a representation of nine seats in the senate is guaranteed to minority parties. The Law of Minorities also reserves 17 seats in the House of Representatives for the party that has lost the election.

Geographically, the island's political organization focuses on municipalities. Puerto Rico is separated into 78 municipalities, each of which is represented by an *alcalde* (mayor), a municipal assembly, and various department heads. The size of the assembly is established according to the size of the population. Members of these assemblies do not receive a fixed salary; instead, they are paid for the meetings in which they participate. Municipalities are arranged around local governments that supervise the general operation of each town. A municipality earns revenue from property taxes and business permits. The revenue is used to pay for school transportation, garbage disposal, local health centers, public works, town police, emergency services, and recreational centers.

CHAPTER 6

Economic Geography

AGRICULTURE

Agriculture was Puerto Rico's most important economic activity until well into the twentieth century. The island's physical geography influenced the location of its agricultural operations. Sugar cane was the crop grown on the flat, fertile coastal lowlands. Tobacco and coffee were the dominant crops in the highlands. Subsistence farmers planted beans, corn, yams, and manioc in both the highlands and the coastal lowlands.

Close economic ties with the United States fostered the expansion and commercialization of the island's farms, which in turn satisfied the North American demand for tropical products. For decades, sugar, the main product for export, was in great demand. During the second half of the twentieth century, however, Puerto Rico's agriculture began to become less competitive. The country was unable to match the prices of other Latin American producers, who were able to supply U.S. markets with cheaper alternatives. Agriculture in Puerto Rico began to decline in importance. Areas formerly under intensive cultivation for sugar cane were transformed into extensive grazing lands for cattle. By the year 2000, agriculture represented only one percent of Puerto Rico's gross national product (GNP). Today, though, two food production activities are once again experiencing growth. These are the production of milk and poultry.

During the second half of the 1990s, only a few Puerto Ricans were employed on farms and less than half of the island's lands were classified as agricultural. Coffee plantations occupied the majority of the small parcels of land that were under cultivation. Nevertheless, the trend is that every year less and less land is devoted to coffee. A series of devastating hurricanes during recent decades caused a sharp decline in coffee production. Plantains were the second most important crop during the 1990s.

Urban expansion is probably chiefly responsible for the loss of agricultural land. Industries, warehouses, and highways now occupy some of Puerto Rico's most potentially productive land. It is very difficult to stop this seemingly irreversible process. Manufacturing and transportation have become critical to the island's economy.

The island's best agricultural soils exist along the most important rivers, where sediments rich in mineral and organic matter are deposited during rainy season floods. Puerto Rico's karst country soils have certain limitations because of their acidity. Nevertheless, the pastures of this area have transformed the karst landscape into one of the island's most important dairy regions. For several decades, the cultivation of pineapples was an important activity in the karst country, although some producers of this fruit have moved their operations to the island's southern coast.

INDUSTRIES

During the first decades of the twentieth century, the production of cigars and the processing of sugar were the island's most important industries. Throughout the 1920s, the U.S. demand for cheap textile products helped Puerto Rico develop local garment industries. Although centered mainly in the western part of the island, the manufacture of clothing was scattered throughout Puerto Rico. This type of industry was criticized for its deplorable working conditions, in which women and children were exploited. Salaries were low and employees were forced to work an average of sixty hours per week.

The shift in the Puerto Rico's economy from one dominated by agriculture to one led by industry began taking place during the 1940s. Several local and global events contributed to this change. During World War II, numerous military facilities were established on the island. In order to support these installations, new roads had to be constructed. Also, port facilities were developed and new airports were built. This modernization of the island's infrastructure between 1939 and 1945 motivated industrialists to take advantage of these new assets. Initially, the government built factories to produce glass bottles, paper products, and cement. Eventually, these government-owned industries were sold to private companies because it was very difficult and costly to maintain them.

Between the late 1940s and early 1950s "Operation Bootstrap" was getting under way. The governments of Puerto Rico and the United States worked together to develop a program aimed to attract American companies. The program offered exemption from U.S. taxes. Because of Puerto Rico's lower cost of living, U.S. companies also were allowed to pay lower salaries to Puerto Rican workers. Further, the island's government offered to construct buildings for any corporation interested in operating on the island. The Commonwealth trained the new mass of industrial workers and established a bank that provided financial aid to anyone interested in starting a manufacturing operation.

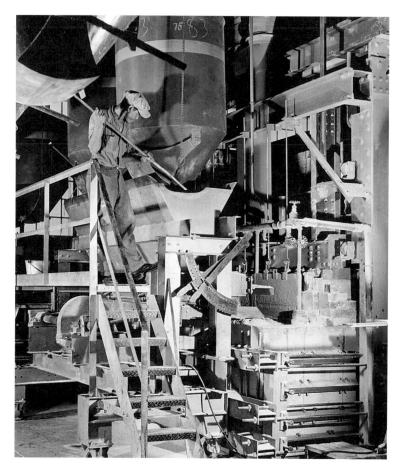

Modernization of Puerto Rico's economic infrastructure became a primary concern in the 1940s and 1950s. During this period, the island's government constructed factories that produced glass bottles, paper products, and cement. Shown here is a factory worker pouring mix into a glass furnace in 1950.

The initial phase of this industrialization program, between 1947 and 1960, stimulated the establishment of light industries that required many workers. It was advantageous to accept light industries on the island, because they did not require a significant amount of initial investment or expensive machinery. Industries that were most likely to prosper during this era were those that involved the manufacture of clothing, shoes, and small electric appliances. Food processing industries also thrived.

During the 1960s, the government's industrialization efforts were aimed at the establishment of petrochemical plants. Several oil refineries were constructed on the northern, eastern, and southern coasts of the island. Industries involved in the manufacture of petroleum derivatives were constructed close to the refineries. After the 1973 global oil crisis, petroleum-based industries lost their importance in Puerto Rico.

Throughout the 1970s, the government invited high-tech industries to establish operations on the island. Unfortunately, however, these highly specialized factories created a limited number of jobs and required very expensive machinery. The salaries offered by these industries were high by Puerto Rican standards, but they required a well-educated staff. Not many Puerto Ricans benefited from the high-tech industries responsible for the manufacturing of electronics and precision instruments. The pharmaceutical industry also came to Puerto Rico during the 1970s. During the 1980s and 1990s, pharmaceutical complexes served as the backbone of the island's manufacturing economy.

The light industries established during the 1950s demanded a significant amount of inexpensive labor. Many of these factories left the island, though, when the minimum wage was raised. When they learned that cheaper labor was available in developing countries, owners of these industries no longer considered Puerto Rico an attractive location for their business. The increase in the island's minimum salaries during the 1960s and 1970s did not affect the relationship between high-tech industries and the Puerto Rican work force, however. It was easier for these industries to adapt to Puerto Rico's rising wages. Problems did begin in 1977, though, when Puerto Rico had to adopt the U.S. minimum wage policy. Before 1977, Puerto Rico had a minimum wage rate much lower than the American minimum salary. This, of course, attracted to the island many industries searching for cheap labor. Nonetheless, labor unions and pro-statehood groups convinced the island's government and the United States to impose American minimum wage laws in Puerto Rico.

From 1976 to the late 1990s, Puerto Rico's economy benefited from Section 936 of the U.S. Internal Revenue Code. Under Section 936, companies that manufacture goods in Puerto Rico were partially exempt from income tax on profits earned from those operations. This complicated loophole, designed to create jobs and relieve Puerto Rico's massive unemployment problem, accounted for 40 percent of the island's $24 billion gross domestic product (GDP) during the mid-1990s. Thanks to Section 936, some 2,000 factories operated on the island during the 1990s, churning out everything from computer floppy disks and underwear to chewing gum and canned tuna—all destined for the huge American market.

Most Washington lawmakers, however, saw the program as a $3-billion-a-year drain on the U.S. Treasury. As a result, in 1996 the U.S. Congress repealed Section 936 of the U.S. Internal Revenue Code, with a clause that retains its benefits for ten years for existing corporations. Many analysts thought that the end of Section 936 would be the end of the island's manufacturing economy. Nevertheless, many industries have decided to stay in Puerto Rico. They are impressed by the solid business environment the island has maintained over the years. Today, the Puerto Rican work force is well educated, thanks to the colleges that serve the island. The main appeal of Puerto Rico is its highly educated and increasingly bilingual work force, its strong infrastructure of roads and utilities, and its strategic geographic location.

THE PHARMACEUTICAL INDUSTRY

Pharmaceuticals are now playing a leading role in the Puerto Rican economy. Some of the top-selling drugs in the

United States are manufactured in Puerto Rico. The island is home to branches of 24 international pharmaceutical companies. In addition, a significant portion of the pacemakers and defibrillators used in the United States are made on the island. By 2004, the pharmaceutical and medical industries were generating more than 60,000 jobs in Puerto Rico. This represents a quarter of the total industrial work force of approximately 240,000. These industries also provide a major market for other enterprises that provide goods and services. Benefactors include people working in such areas as manufacturing components, food services, engineering, medical and legal services, banking services, insurance, transportation, communications, tourism, and others.

THE CONSTRUCTION INDUSTRY

The construction industry is one of the most important activities of the island's manufacturing sector. Since the 1940s, the construction of buildings, factories, homes, ports, and roads has been an important contributor to Puerto Rico's economy. Several factors combined to create conditions favorable to the emergence of large construction companies. These conditions include the island's growing population and industrialization, and the implementation of government programs that made houses affordable. During the 1950s, such companies made significant profits from highway construction projects and the development of residential complexes.

Throughout the 1990s, the island's construction industry experienced additional growth when the government engaged in the modernization of the island's infrastructure. During that era, a super aqueduct was constructed to deliver water from the west central mountains to the crowded city of San Juan. This aqueduct became necessary when accumulated sediments from nearby urbanized areas caused the reservoirs close to the capital to lose their water storage capacity. Another factor was behind the decision to build the super aqueduct:

The island's economy suffered greatly during the severe drought of 1994. Today, the super aqueduct distributes up to 75 million gallons of water daily to 15 municipalities along the northern coast and benefits 1.6 million islanders. This mega-project was completed in 2000.

Another project that benefited the island's construction companies was the building of a commuter railroad in the San Juan metropolitan area. Puerto Rico's commuter train was built to reduce highway congestion and to provide rapid transportation for the 1.4 million people of the crowded capital area. The train was the first of its type in the West Indies.

During the past half century, the construction industry has been responsible for transforming some of the island's green scenery into urban landscapes. Financial backers of the construction companies, as well as the companies themselves, have been blamed, though, for the deterioration of Puerto Rico's environment. Beyond question, infrastructure development projects have negatively affected sensitive coastal environments, mountain forests, wetlands, and swamps. Not everyone agrees on who is to blame, however. Some analysts accuse the Puerto Rican government. It, after all, granted the building permits to developers who have little concern for the natural environment and then failed to regulate the process.

SERVICE INDUSTRIES

The service sector of the Puerto Rican economy is growing rapidly. Service-related industries include banking, real estate agencies, and insurance corporations. Also included are travel and tourism companies, transportation services, educational institutions, health care, and legal services. Puerto Rico is the banking center of the West Indies. In addition to its own banks, most large U.S. and many European banking corporations have branch offices in Puerto Rico. During recent years, the island's service sector has experienced more rapid growth than have the non-service-related industries. The government is Puerto

Economic Geography

More than one million U.S. tourists visit Puerto Rico each year, helping to generate billions of dollars for Puerto Rico's economy. Tourists are shown here relaxing at a beachfront resort.

Rico's chief employer. In fact, one of every five people in the island's work force is a government employee.

Tourism has made an important contribution to the island's economy. Every year, about one million Americans visit Puerto Rico. They come to enjoy the perpetual summer weather, far-reaching sand beaches, exotic culture, and diverse tropical landscapes. Puerto Rico offers a fabulous array of activities for those who enjoy water sports, hiking, golf, and tennis. The island is home to some of the most beautiful beaches in the West Indies—varying from pure white dunes in the northwest to the black volcanic sands of the southeast. There are hundreds of guesthouses and more than 30 large resorts, each characterized by grand swimming pools, excellent facilities for tennis, and long stretches of scenic beachfront. The largest hotels have casinos and beautifully groomed golf courses. The government regulates casinos carefully. One of the countless reasons why many Americans select Puerto Rico as their Caribbean destination is because no visa or passport is required for U.S. citizens entering the island. In addition, all business in Puerto Rico is transacted with U.S. dollars. Finally, the island has no sales tax, and the island's mail is administered by the U.S. Postal Service.

The capital city's historical district, known as Old San Juan, continuously attracts island visitors. This quaint area is a place of narrow cobblestone streets, colorful colonial buildings, centuries-old fortresses overlooking the Atlantic Ocean, and fascinating museums and art galleries. The district was founded in 1521, thereby making it the oldest European city under the U.S. flag. Old San Juan has more than 400 carefully renovated sixteenth- and seventeenth-century Spanish colonial houses, commercial buildings, and military structures. The National Park Service of the U.S. Department of the Interior maintains two imposing fortresses rising dozens of feet above the sea. In addition, Old San Juan boasts two colonial churches and a gothic cathedral.

CHAPTER 7

Living in Puerto Rico Today

Puerto Ricans are recognized for their generosity toward island visitors. Islanders take equal pride in their U.S. citizenship and in their Hispanic and Caribbean culture. Although U.S. products and cultural influence are pervasive, many in Puerto Rico defend the island's traditions against what they see as an intrusion of foreign culture. The Puerto Rican Institute of Culture, created in 1955, has played a vital role in preserving many aspects of the island's culture. The institute's mission includes the conservation, enrichment, and diffusion of the island's cultural treasures. In addition, it educates Puerto Ricans about their heritage and creates in them a sense of pride in their cultural traditions. The agency is engaged in a wide variety of programs that include supporting graphic arts, sculpture, and popular arts exhibitions. It also supports archeological investigations, management of museums and parks, and maintenance of monuments and historic districts and celebrates Puerto Rican music, dance, and theater.

THE ARTS

Puerto Ricans are well-known in Latin America for their creativity in the graphic arts. During the eighteenth and nineteenth centuries, painters such as José Campeche (1751–1809) and Francisco Oller (1833–1917) brought fame to Puerto Rico. Campeche's work exhibited superb composition, precision in drawing, and beauty of color. His most remarkable paintings are "A Lady Riding a Horse," "Our Lady of Mercy," and "Our Lady of Carmen." Oller's portraits of government officials and workers and landscapes of sugar plantations and peasant shacks celebrated both the island's natural beauty and its social problems. "The Wake" is Oller's most famous painting. This painting shows the wake of a peasant child. During the nineteenth century, it was a tradition among the peasants to celebrate the death of a child. They thought that a dead child would become an angel, a protector of the community. The people in the scene are partying while the parents suffer their loss. Oller considered this tradition absurd, and the painting was a protest.

Famous Puerto Rican painters of the twentieth century include Francisco Rodón and Lorenzo Homar. Rodón's paintings have been auctioned by New York's prestigious Sotheby's Auction House. His work includes portraits of important Latin American cultural figures. Homar is considered the most influential artist of the past 60 years. Not only does Homar paint, but he is also a famous engraver and printmaker. He is credited as the artist most responsible for promoting printmaking in Puerto Rico, and he trained other famous Puerto Rican painters.

Puerto Rican writers have been praised for their tendency to write brilliantly about the everyday events that affect the lives of the island's populace. Novels produced during the twentieth century describe the problems of identity those native to the Caribbean islands have because of the islands' proximity to the United States. The works of Luis Rafael Sánchez, the most famous writer of the last three decades of the twentieth century, have been translated into a number of languages. He is a prolific essayist, playwright, and novelist. Another important Puerto Rican writer is Enrique Laguerre. The hallmark of Laguerre's writings is his detailed descriptions of the social problems that affect the rural and urban areas of the island. He was nominated for the Nobel Award in Literature during the late 1990s.

Even though Puerto Rico is a small Latin American island, it has given birth to diverse forms of folk music. Music genres include bomba, plena, danza, and décima. Bomba music has strong African roots, and percussion instruments are very important in every performance. Descendants of African slaves played a crucial role in the development of this form of music—African dances are associated with bomba music.

Bomba music came from the rural plantations, but Puerto Rican plena was created in the island's southern cities. Plena is considered more eclectic than bomba because it involves the use of European and African instruments and rhythms, whereas bomba is decisively African. Plena music reflects the united Spanish and African heritage of islanders. Plena musicians employ guitars, accordions, tambourines, and brass instruments. Plena's lyrics are linked to contemporary happenings and are often considered as a kind of living newspaper. Performers sing about current events and often ridicule local politicians or noteworthy occurrences.

Danza music traces its origins to Spanish waltzes. Originally, the danza was dance music, played by instrumental groups or a solo piano. This type of musical expression hailed from the island's elite during the second half of the nineteenth and first half of the twentieth centuries. In essence, the danza was the main form of entertainment for Puerto Rico's aristocracy. As the form developed, words were set to existing melodies and melodies developed for well-known poems. A string orchestra, woodwinds, and an elegant atmosphere characterize this form of Puerto Rican music. Danza music gained popularity

Puerto Rico is known for its diverse genres of music, including bomba, plena, danza, and decima. Shown here are dancers performing to plena, a form of folk music in which the lyrics are tied to current events.

when young musicians wrote songs with lyrics that celebrated the national identity of Puerto Ricans and their love for their country.

The decima is perhaps the most alluring type of Puerto Rican folk music. It is the vehicle through which the traditional people of the hills of Puerto Rico communicate happiness and sadness. Instrumentation for the decima consists of a Puerto Rican cuatro (a ten-stringed instrument similar to a small guitar and considered the island's national instrument), and a guiro. This instrument is made of the dry, hard shell of a cucumber-like fruit and is played by scraping the surface with a stick. Decima artists sing about love, important events, the human condition and experiences, the Puerto Rican people, or individuals to whom the performer wants to show respect.

The trademark of the decima is verbal improvisation. Decima improvisation competitions are held in many rural areas of the island.

SOCIAL PROBLEMS

Many Puerto Ricans, particularly those living in large urban centers, live in constant fear because of the high incidence of violent crimes and crimes against property. Each day, local newspapers and television news programs stress the seriousness of such delinquent acts as murders, burglaries, robberies, domestic abuse, corporate crimes, and government corruption. Crime is considered to be the island's leading problem. From 2000 to 2004, Puerto Rico's murder rates were higher than in any U.S. state. With 790 murders recorded during 2004, Puerto Rico had a higher murder rate than the U.S. cities of New York, Los Angeles, and Chicago, even though Puerto Rico has a smaller population. The island's murder rate is three times that of the United States and four times that of European countries.

It is estimated that between 70 percent and 80 percent of all the crimes committed during the 1990s and early 2000s can be linked to drug dealing. Some murders occurred when drug dealers fought among themselves to protect "their sales territories." Others are the execution of clients who owe dealers money. In addition, some murders have been linked to drug addicts who kill their robbery victims. Drug dealing is an illegal activity that has greatly affected the island's social stability and the health of its population. Drug dealers are making huge sums of money, but drug addicts have seen their health deteriorate and their families destroyed. Drug dealing is a tragic problem that affects the entire island. Serious problems of drug addiction began when Puerto Rican addicts living in New York returned to the island in the 1950s and 1970s. Before that, the islanders were not familiar with the drugs used in U.S. urban ghettoes. Many Puerto Ricans who came to America in search of a better life encountered only frustration, poverty, and social disorientation. Some turned to drugs. When they returned to the island, they imported the vice.

Puerto Rico's strategic geographic location and its close political association with the United States has transformed the territory into an important intermediary transportation point for those who are trying to export illicit drugs from South America to the United States. Puerto Rico lies closer than Miami to the Colombian harbors from which many drugs are exported. Once the drugs reach Puerto Rico, dealers do not need to worry about interference from U.S. border patrols, customs, or the U.S. Drug Enforcement Agency. Therefore, little risk is involved in transporting drugs from the San Juan International Airport or the island's harbors to the mainland.

The governments of Puerto Rico and the United States have collaborated closely to find ways to stop the flow of drugs onto the island. One strategy involves the development of radar systems in the southern part of Puerto Rico to detect the approach of planes or boats carrying illegal drugs. Both governments are investing heavily in rigid law enforcement programs to stop illegal drugs from entering the country. Some critics believe that the huge sums of money used to support border patrol programs could be used more effectively to treat addicts. They believe that the social work approach is less costly and more helpful than the rigid law enforcement approach. The social work approach does not perceive the drug addict as a criminal, but as the victim of a disease.

Social and economic inequalities are considered a serious problem in Puerto Rico, and they probably contribute to the country's drug abuse problem. Almost 50 percent of the island's population lives below the poverty level. In Central America's most stable nation, Costa Rica, only 18 percent of the population lived below the poverty line in 2004. Although Puerto Rico has experienced a gradual capitalist transformation into a high-tech manufacturing center, it is an unfortunate reality that the island is full of contradictions and inequalities.

Living in Puerto Rico Today

San Juan's dynamic landscape is one of skyscrapers housing financial institutions, luxurious hotels, and many other thriving businesses. This impression can be deceiving if the visitor does not take into account that beyond that skyline of modern buildings there is another Puerto Rico dominated by stagnation. Many Puerto Rican families still do not have access to adequate housing. And about half of the island's adult population has not been able to finish high school, even though literacy rates are among the highest in Latin America. The high rates of school dropouts can lead to increased levels of unemployment. During recent years, Puerto Rico's unemployment rate has hovered around 12 percent, whereas in the United States it has been about 5.5 percent. Costa Rica, which is the only Middle American country comparable to Puerto Rico in terms of population size and political stability, had an unemployment rate of 6.6 percent.

Puerto Rico Looks Ahead

Puerto Rico's social and economic challenges will continue to pose a formidable challenge well into the future. On the positive side, the island is a magnet for tourists and high-tech corporations because of its physical and human geography. Nonetheless, the territory faces many serious problems. Puerto Rico has become one of the main points of U.S. entry for the South American illegal drug trade. According to estimates made by the U.S. Drug Enforcement Administration, up to 20 percent of all the cocaine entering the United States comes through Puerto Rico or the U.S. Virgin Islands. Additionally, the island's crime rates are much higher than those in the United States. Politically, uncertainty is the norm. Neither the U.S. Congress nor the people of Puerto Rico seem able to decide on the island's political future—whether to retain commonwealth status, seek independence, or become the fifty-first U.S. state.

Puerto Rico Looks Ahead

Anibal Acevedo Vila waves the flag of Puerto Rico shortly after he became the eighth governor in January 2005. Vila is shown here with his children, Gabriela and Juan Carlos, and his wife, Luisa.

The island also faces serious environmental challenges. Of major concern is the preservation of an ample supply of clean, fresh water to meet the needs of its growing population. Islanders must preserve the quality of renewable environmental resources for future generations to use and enjoy. Even though numerous creeks and small rivers flow from the rainy interior highlands to the island's reservoirs and aquifers, water shortages were common during the 1980s and 1990s. The destruction of forests and the elimination of vegetation along the riverbanks have limited water absorption cycles. In addition, deforestation has been responsible for the accumulation of sediments in river channels and reservoirs. This increases the chance of flooding and limits the amount of water that can be stored. Since the island is so heavily populated, some individuals without access to landfills have thrown garbage in rivers and creeks, adding to their pollution. Industries and poorly managed water treatment plants have also contributed pollutants to Puerto Rico's waterways. Further, some manufacturing operations have dumped toxic wastes into streams.

Some Puerto Ricans are not environmentally conscious. Sometimes, there is no significant participation of the island's population in programs dedicated to the preservation of water resources. Rapid urbanization is yet another reality that places pressure on the dwindling supply of fresh water. Development of extensive housing projects puts great pressure on existing supplies. Finally, a significant amount of water from the island's reservoirs is lost because of poorly maintained water delivery systems. Broken pipes are common; 40 percent of the drinking water transported by distribution lines is lost because some part of the system is not functioning properly.

Puerto Rico's electricity is produced by burning fossil fuels, a method that contributes to atmospheric pollution. Areas close to thermal-electric facilities have reported air temperatures higher than those in nearby rural areas. Puerto Rico's environment has also suffered considerably because of poor solid waste management. The island's urban areas do not have adequate space to accommodate landfills. For this reason, some islanders have created illegal dump sites that do not follow local and federal guidelines for the proper disposal of solid waste. Not only do such sites add to environmental pollution, they also contribute to visual blight of the island's once beautiful rural landscapes. If islanders want to attract more tourists to this tropical paradise, they must do a much better job of safeguarding their environment's tropical landscapes and natural resources.

Northern Puerto Rico has experienced rapid population growth and explosive urbanization. Despite having compromised the quality of the island's environment, this growth has been beneficial to American and Puerto Rican corporations. Puerto Rico's urban population is a main consumer of American manufactured goods and foodstuffs. There are numerous positive commercial links between the United States and Puerto Rico's urban centers. For example, San Juan is now the largest market in the Caribbean region. As such, a significant number of American and multinational corporations have established regional headquarters on the island. San Juan's international airport and harbor have expanded to satisfy the needs of the capital's growing population. At the same time, these transportation facilities have converted the San Juan metropolitan area into the Caribbean's main transportation hub.

If American industries, transportation companies, and banks were to leave Puerto Rico, surely the island's economy would suffer a sharp downturn. For this reason, continued governmental cooperation between the United States and Puerto Rico, including incentives to corporations that operate on the island (or may elect to do so in the future), is necessary.

The island is blessed with talented people and an urban market that is loyal to American products. Politically, however, it remains a commonwealth with many important decisions still imposed by the United States. If Puerto Rico is going to continue prospering, the United States must assist the island in the continuation of its modernization process. The author, himself a native-born Puerto Rican, is confident that the United States and Puerto Rico—working together—have the capacity to successfully guide the island's the environmental, social, and economic future.

Facts at a Glance

Land and People

Official Name Commonwealth of Puerto Rico.

Conventional Name Puerto Rico

Capital San Juan

Location The island is the most easterly of the Greater Antilles group of the West Indies. Puerto Rico is bounded on the north by the Atlantic Ocean, on the east by the Virgin Islands Passage, on the south by the Caribbean Sea, and on the west by the Mona Passage, which separates the island from the Dominican Republic. The Caribbean coast of Venezuela is 597 miles (960 kilometers) south of Puerto Rico, and the distance between the island and New York City is 1,491 miles (2,400 kilometers).

Area The area of the island is 3,423 square miles (8,866 square kilometers). Puerto Rico is only 111 miles (185 kilometers) long by 39 miles (63 kilometers) wide, about three times the size of the state of Rhode Island.

Climate The island's climate is tropical, but within this broad definition there are variations. The north coast and the mountainous interior are rainy; the southern lowlands relatively dry. The average annual temperature is 79°F (26°C), with winter and summer averages of 76 (24) and 82 (28) degrees, respectively. Geographers call Puerto Rico's climate a tropical maritime climate.

Terrain Puerto Rico is a mountainous tropical island with several distinct geographic regions. About three-fourths of its territory is hilly or mountainous. The island can be divided into three regions, based on landform: interior highlands, coastal lowlands, and northern limestone karst country.

Elevation Extremes Lowest point: the Caribbean Sea, at 0 feet. Highest point: Cerro Punta, at 4,389 feet (1,338 meters).

People	Nationality: Puerto Ricans are U.S. citizens.
Population	3,916,632 (July 2005 estimate).
Population Density	1,100 per square mile (437 per square kilometer). The most heavily populated municipality in the island is its capital city, San Juan, with a population density of 9,328 inhabitants per square mile (3,602 per square kilometer). Only in remote towns between the lush hills of central Puerto Rico are there population densities as low as 400 persons per square mile.
Population Growth Rate	0.47 percent (2005 estimate). World average: 1.3 percent.
Fertility Rate	Average number of births per woman in her childbearing years: 1.91 (2005 estimate).
Life Expectancy	Total population: 77.62 years. Male: 73.67 years. Female: 81.77 years (2005 estimate).
Median Age	Total: 34.23 years. Male: 32.5 years. Female: 35.87 years (2005 estimate).
Literacy	Definition: Age 15 and over can read and write. Total population: 94.1 percent. Male: 93.7 percent. Female: 94.4 percent (2002).
Ethnic Groups	Majority of mixed African and Spanish descent. Black, 11 percent. Amerindian, 0.4 percent. Asian, 0.2 percent.
Languages	Spanish, English.
Religions	Roman Catholic, 67.5 percent; Protestant groups, 32.5 percent.
Economy	
Land Use	Arable land, 3.95 percent. Land in crops, 5.52 percent. Other, 90.53 percent.
Irrigated Land	240 square miles (400 square kilometers; 1998 estimates).
Natural Hazards	Periodic droughts, hurricanes, landslides; seismic and tsunami hazards.
Environmental Issues	Deforestation, erosion; occasional drought, causing water shortages.

Facts at a Glance

Geography Note	Important location along the Mona Passage (a key shipping lane to the Panama Canal). San Juan is one of the biggest and best natural harbors in the Caribbean. Many small rivers and high central mountains ensure land is well watered. The southern coast is relatively dry. The northern part of the island is a fertile coastal plain belt.

Government

Type of Government	Commonwealth associated with the United States.
Head of State	President of the United States. Local administration of the territory and executive power is vested in a governor, elected to a four-year term.
Independence	None (commonwealth associated with the United States).
Administrative Divisions	78 municipalities.
Currency	U.S. dollar.
Gross Domestic Product	Estimated to be $69 billion (2004 estimate).
Labor Force	1.3 million (2000).
Unemployment	12 percent (2002).
Labor Force by Occupation	Agriculture, 3 percent. Industry, 20 percent. Services, 77 percent (2000 estimate).
Industries	Pharmaceuticals, electronics, construction, apparel, food products, tourism.
Agricultural products	Sugarcane, coffee, pineapples, plantains, bananas, livestock products, chickens.
Exports	$46.9 billion (2001).
Export Commodities	Chemicals, electronics, apparel, canned tuna, rum, beverage concentrates, medical equipment.
Export Partners	United States, 90.3 percent. Great Britain, 1.6 percent. Netherlands, 1.4 percent. Dominican Republic, 1.4 percent (2002 estimate).
Imports	$29.1 billion (2001).

Import Commodities	Chemicals, machinery and equipment, clothing, food, fish, petroleum products.
Import Partners	United States, 55.0 percent. Japan, 5.4 percent (2002 estimate).
Transportation	Highways, total: (15,738 miles, or 25,328 kilometers). Paved: 14,705 miles (23,665 kilometers, including 265 miles, or 426 kilometers, of expressways). Unpaved: 847 miles (1,363 kilometers; 2004). Railroads total: 60 miles (96 kilometers). Narrow-gauge railroads: 60 miles (96 kilometers). Ports and harbors: Aguadilla, Arecibo, Fajardo, Guanica, Guayanilla, Guayama, Mayaguez, Playa de Ponce, San Juan. Airports: 30 (2004 estimate). Airports with paved runways: 17.
International Issues	Increasing numbers of illegal migrants from the Dominican Republic cross the Mona Passage to Puerto Rico each year looking for work.
Note	Many of the data contained in this summary were taken from *CIA–The World Factbook, Puerto Rico (2005)*. For updated data, see: *www.cia.gov/cia/publications/factbook* (check "Puerto Rico").

History at a Glance

2000 B.C.	The earliest inhabitants of Puerto Rico were Amerindian groups that were part of different migratory waves. Some anthropologists believe that the first natives came from North America.
A.D. 100	Arawak Indians migrate to Puerto Rico from South America.
1493	The European explorer and navigator Christopher Columbus sighted Puerto Rico.
1500s	Europeans living in Puerto Rico and surrounding territories realized that sugar cane was their most lucrative cash crop. Sugar enjoyed great demand in Europe.
1508	Spanish conquerors arrived to colonize the island and responded to the Tainos' naïve welcome with cruelty and greed once they learned of the presence of gold in the territory.
1521	The city of Old San Juan is founded.
1530	The Amerindian population dwindled to 1,553 because the majority of the native population died from diseases imported by the Europeans and also because they were mercilessly overworked in the mines and agricultural fields. Because of their large population, African slaves became the dominant group.
1500s–1700s	The construction of fortresses and defensive walls consumed most of the island's limited resources. When these military projects were completed, thick walls surrounded the capital city of San Juan, and only a few gates allowed the residents of the walled city to leave.
Late 1600s	The island's sugar industry was unable to compete with producers from other islands of the Caribbean. For this reason, the rural geography of the island changed from one dominated by sugar plantations to a more diverse landscape in which cattle ranches gained importance.
1700s	Outside the capital city, the owners of cattle ranches were the most powerful group in the countryside.
1787	The island's population was 103,051, and the number of slaves 11,260, representing 11 percent of the colony's population.
Early 1800s	Spain lost the majority of its possessions in the Americas.
1809–1868	Spanish authorities discovered eight conspiracies organized to start uprisings that would lead the island to independence.

1846 The slave population represented only 11.6 percent of the island's population.

1868 Several hundred men took a town in the central highlands of Puerto Rico, arrested its officials, and proclaimed the Republic of Puerto Rico. The new independent country would be short lived, however. As soon as the Spanish authorities found out what had happened, they mobilized their forces and marched into the island's interior. They achieved an easy victory over the revolutionaries.

1873 The first Protestant church was dedicated in Puerto Rico. This place of worship was an Anglican church.

1873 Slavery was abolished.

1898 The Spanish government granted autonomy to the island, the first system of home rule the island had ever known. The island enjoyed a very brief period of autonomy, however, because U.S. naval forces moved to occupy the territory in July 1898. Once Spain was defeated by the United States during the Spanish-American War, Puerto Rico became an American territory.

1899 The most devastating hurricane in Puerto Rico's history, San Ciriaco, was responsible for the deaths of 3,369 people.

1903 The University of Puerto Rico was founded.

1917 After 18 years of U.S. occupation, the people of Puerto Rico finally attained American citizenship and to some extent a greater degree of self-government.

1918 The island of Puerto Rico was struck by an earthquake of magnitude 7.5, centered approximately 9 miles (15 kilometers) off the island's northwestern coast. This earthquake produced a tsunami that struck several towns on the island's west coast. Of the 116 people killed during this seismic event, 40 were victims of the tsunami.

1932 During local elections in Puerto Rico, women voted for the first time, although illiterate women were not allowed to vote until 1935.

Late 1930s Progressive autonomist leader Luís Munoz Marín founded the Popular Democratic Party.

Early 1940s Many Puerto Ricans lost their lands, as the American military expropriated sections of the island for the establishment of military facilities aimed to defend the Caribbean Sea and the Middle Atlantic from possible German attacks.

History at a Glance

1952 The United States allowed Puerto Rico to have local home rule through popular election of a governor and members of a Senate and a House of Representatives. Before 1952, the president of the United States appointed the island's governors. The territory's administration system, established in 1852 and known as the Commonwealth, has some limitations, though, because it does not allow islanders to vote in American presidential elections or to send voting representatives to the United States Congress.

1956 The contribution of the industrial sector to the gross production of goods and services was more than the contributions of farms and plantations.

1957 With the assistance of the United States, the Commonwealth founded the Housing and Urban Renewal Corporation. This agency built more than 34,000 public housing units during the 1960s.

1960 Government industrialization efforts were aimed at the establishment of petrochemical plants.

1964 The island's most important healthcare facility, the Puerto Rico Medical Center, was opened to the general public.

Late 1960s The island's pro-statehood movement gained importance, and the New Progressive Party was founded. The main goal of this political party is to make Puerto Rico the fifty-first state of the American Union, so islanders can enjoy the same rights and privileges that benefit U.S. citizens.

1967 The island held a referendum, and those who favored statehood represented only 38.9 percent of the vote.

1970s The government of Puerto Rico invited high-tech industries to establish operations on the island. These factories encountered problems because of the limited number of jobs they created and their demand for very expensive machinery.

1976 to the late 1990s Puerto Rico's economy benefited from Section 936 of the U.S. Internal Revenue Code. Under Section 936, companies that manufacture goods in Puerto Rico were partially exempt from income tax on profits earned from those operations.

1977 Puerto Rico had to adopt the U.S. minimum wage policy.

1980s and 1990s   Pharmaceutical companies were the backbone of the island's manufacturing economy.

1990s   In two referendums, Puerto Ricans had the opportunity to vote if indicate if they were in favor of statehood, independence, or the commonwealth status, and on both occasions, the commonwealth option prevailed.

1998   Hurricane Georges crossed the island from east to west with merciless winds that struck the territory at great length.

2000   Agriculture represented only 1 percent of the national gross product. The super aqueduct project was completed. This project was designed to deliver water from the west central mountains to the overcrowded city of San Juan.

2001   32.5 percent of the island's population belonged to Protestant religious groups (mainly Baptists, Episcopalians, Lutherans, Methodists, Pentacostals, Disciples of Christ, and Presbyterians).

Further Reading

Blouet, Brian W. and Olwyn M. Blouet. *Latin America and the Caribbean: A Systematic and Regional Survey.* Hoboken, NJ: Wiley, 2001.

Castro-Arroyo, Maria and Luque-de-Sanchez, Maria. *Puerto Rico En Su Historia: El Rescate de la Memoria.* Rio Píedras, PR: Editorial La Biblioteca, 2001.

Clawson, David. *Latin America & the Caribbean: Lands and People.* Dubuque, IA: McGraw-Hill, 2005.

Cohen, Tina. *Off the Beaten Path: Puerto Rico,* 4th ed. Guilford, CT: Insider's Guide, 2005.

Cruz-Baez, Angel, Carlos Guilbe and Adolfo Lopez. *Vive la Geografia de Nuestro Puerto Rico.* San Juan, PR: Editorial Cordillera, 2002.

Hawn, A. Carleen. *Puerto Rico,* English ed. Florence, Italy: Bonechi, 2004.

Luxner, Larry. *Insight Pocket Guide: Puerto Rico.* Singapore: APA Publications, 2005.

Miner-Sola, Edwin. *Endangered and Threatened Species of Puerto Rico.* San Juan, PR: Ecologic Series, 2005.

Peffer, Randall. *Puerto Rico,* 2nd ed. Melbourne, Australia: Lonely Planet Publications, 2002.

Pitzer, Kurt, and Tara Stevens. *Adventure Guide to Puerto Rico,* 4th ed. Edison, NJ: Hunter Publishing, 2001.

Porter, Darwin, and Danforth Prince. *Portable Puerto Rico.* Hoboken, NJ: Wiley Publishing, 2005.

Rodriguez-Marcos, Julian. *Puerto Rico,* English-Spanish ed. La Coruna, Spain: Everest Editorial, 2004.

Stallings, Douglas, ed. *Puerto Rico,* 3rd ed. New York: Fodor's Travel Publications, 2005.

West, Robert C. and John P. Augelli. *Middle America: Its Lands and Peoples.* Upper Saddle River, NJ: Prentice Hall, 1989.

Websites

CIA—The World Factbook, Puerto Rico
http://www.cia.gov/cia/publications/factbook

Puerto Rico Tourism Company
http://www.gotopuertorico.com/

Puerto Rico Government Websites
http://www.govengine.com/stategov/puertorico.html

General information on Puerto Rico
http://www.puertoricowow.com/html/PuertoRico_facts.htm

Internet Guide to Puerto Rico
http://www.topuertorico.com/

Exploring Puerto Rico
http://welcome.topuertorico.org/

Index

Abolitionist movement, 38, 40
Africans, 50–52
African slaves, 34, 37–38, 50
Agriculture, 42, 71–72
Aguadilla, 27
Alomar, Roberto, 11
Amerindians, 32, 49
Anglican churches, 55
Animal and plant life, 27–30
Aquifers, 22
The Arts, 82–85
Auditor, The, 59
Ausubo trees, 28
Automobiles, 47

Ball courts, ceremonial (*bateyes*), 50
Bateyes (ceremonial ball courts), 50
Beaches, 24, 79–80
Bilingualism, 76
Bioluminescence, 19
Birds, 28
Birth rates, 49–50
Blending of three races, 52–53
Borinquen, 32
Botanicas, 53

Camuy River, 21
Caribbean National Forest (*Il Yunque*), 9, 16–17
Cassava (manioc) fruits, 50
Castillo de San Felipe del Morro, 36
Caves, 21
Central highlands, 16–17
Ceremonial ball courts (*bateyes*), 50
Charter (1952), 70
Class distinctions, 36–37
Clemente, Roberto, 11

Climate, 22–24
Coastal lowlands, 18–19
Cocaine, 88
   See also Drugs, illegal
Cockfights, 10–11
Coffee, 71–72
Cold War, 61
Colonizers, Spanish, 27–28
Columbus, Christopher, 32
Commonwealth, birth of, 62–64
Commonwealth Senate, 70
Copper, 31
Coqui (tree frog), 29–30
Coral reefs, 29–30
Cordero, Angel "Junior," 11
Cordillera Central mountain range, 16
Cortes (National Assembly), 39
Creoles, 36
Crimes, high incidence of, 85
Culture. *See* People and culture

Deforestation, 25–26, 89
del Toro, Benicio, 11
Department of Health, 67
Dinoflagellates, 19
Disease, 34, 49
Dominican Republic, 11
Drugs, illegal, 85–86, 88

Earthquakes, 26–27
Economic geography, 71–80
   agriculture, 42, 71–72
   construction industry, 77–78
   industries, 73–76
   pharmaceutical industry, 76–77
   service industries, 78–80
Economic inequalities, 86–87
Education, 64–66
Encomienda system, 33–34
English language, 54, 69

Index

Environmental mismanagement, 90
Europeans, 52

Farming, subsistence, 71
Feliciano, Jose, 11
Fernandez, Gigi, 11
Fishes, 29
Flamboyant (flame tree), 28
Flame tree (*flamboyant*), 28
Flooding, 26
Forest types, 17
Free associated state, 62
Future of Puerto Rico, 88–91

Geography, economic. *See* Economic geography
Geography, political. *See* Political geography
Georges, Hurricane, 25
Gibraltar of the Caribbean, 61
Giralt, Ramon Power, 39
Gold, 33–34
Governor, 70
Guanica, 41
Guaraguao (red-tailed hawk), 28

Haiti, 11
Health, 66–68
Highlands, central, 16–17
Highways, 46–47
Higuera trees, 28
Hispaniola, 11
Historical geography, 32–45
    independence movements, 39–42
    modernization of the island, 43–45
    overview, 32–33
    politics in last half of 20th century, 45

    social classes in colonial era, 35–38
    Spanish conquest, 33–35
    Spanish control, 38–39
    United States and, 42–43
Home rule, 41, 42
House of Delegates, 57
House of Representatives, 70
Housing and Urban Renewal Corporation, 63–64
Hurricanes, 24–26

Illiteracy, reduction of, 64, 87
Independence movements, 39–42
Industries, 73–80
    construction, 77–78
    overview, 73–76
    pharmaceutical, 76–77
    service, 78–80
Inequalities, economic, 86–87
Inter-American University, 66
Intermarriage among races, 49
Internal Revenue Code, Section 936 (U.S.), 76

Jones Act, 58–59
Jutias, 29

Karst country, 20–22

Lagoons, mangrove, 19
Language, 52, 53–54
La Parguera, 20
La Plata River, 24
Law of Minorities, 70
Life expectancy rates, 67
Literacy, 64, 87
Loiza, 50–52
Lowlands
    coastal, 18–19
    northern, 18, 24

Index

Maine incident, 57
Manatees, 28–29
Mangrove swamps, 19–20
Manioc (cassava) fruits, 50
Maracas, 28
Marin, Luis Munoz, 45, 59–60, 62
Martin, Ricky, 11
Mayors (*alcaldes*), 70
Medical Center, Puerto Rico, 67
Medicine, School of, 68
Mestizo (mixed ancestry), 53
Minerals, 30–31
Minimum wage laws, 74–75
Mining, 31
Minorities, Law of, 70
Modernization of the island, 43–45, 73–74
Mogotes (karst hillocks), 20–22, 21
Mortality rates, 67
Mulattoes, 53
Municipalities, 70
Music, 50, 53

National Assembly (*Cortes*), 39
Native groups, 49–50
Northern lowlands, 18, 24

Occupation, U.S., 57–61
Oil refineries, 75
Old San Juan, 8, 36, 80
Operation Bootstrap, 44, 45, 63, 73
Ortiz, Piculin, 11

Palo colorado forest, 17
Palo de pollo tree, 28
Panama Canal, 11–13

Parrot, Puerto Rican, 17, 28
People and culture, 46–54
    Africans, 50–52
    blending of three races, 52–53
    Europeans, 52
    language, 53–54
    native groups, 49–50
    overview, 46–49
    religion, 54–55
Pharmaceutical industry, 75
Pineapples, 72
Pitirre birds, 28
Plant and animal life, 27–30
Plates, tectonic, 26
Political geography, 56–70, 57
    birth of commonwealth, 62–64
    education, 64–66
    health, 66–68
    overview, 56–57
    political organization, 70
    pro-statehood movement, 68–69
    U.S. occupation, 57–61
Political organization, 70
Political status, 42, 45
Politics in last half of 20th century, 45
Pollutants in water system, 89–90
Pollution, 22
Ponce de Leon, Juan, 33
Popular Democratic Party, 59–60
Population growth, 46–49, 52
Privateers, 35
Pro-independence movement, 59
Pro-statehood movement, 68–69
Protestant churches, 55
Public housing projects, 63–64
Puerto Rican Suffrage League, 59–60
Puerto Rico
    as American territory, 40–43
    celebrities of, 11

Index

climate of, 22–24
culture of, 9–10
earthquakes and, 26–27
future of, 88–91
geography of, 9–11
hurricanes and, 24–26
minerals of, 30–31
people of, 9–10
population of, 46–49, 52
statehood options of, 45
strategic location of, 34–35, 58–59, 61
topography of, 14–16
United States and, 9, 11–13, 41–45, 57–58, 90–91

Racial blending, 52–53
Rainfall, 17, 22, 23–24
Red-tailed hawk (*guaraguao*), 28
Reefs, coral, 29–30
Refineries, oil, 75
Religion, 53, 54–55
Reptiles, 29
Reservoirs (aquifers), 22
Rio Píedras, 64–66, 67
Rodriguez, Chi Chi, 11
Rodriguez, Ivan, 11
Roman Catholic Church, 54–55
Roman Catholicism, 52

Sanitary conditions, improvement of, 63
San Juan, 8–9, 11–13, 38, 46
  See also Old San Juan
San Juan International Airport, 13
Santeria, 53
Santiago Apostol Festival, 51
School of Medicine, 68
Separatist groups, 40
Shantytowns, 64
Sierra de Cayey, 17

Sierra de Luquillo, 16–17, 24
Sinkholes, 16, 21–22
Slaves, African, 34, 37–38, 50
Social classes in colonial era, 35–38
Social problems, 85–87
Southern lowland belt, 18–19
Spain, 38, 41–42, 48
Spanglish, 54
Spanish-American War (1898), 9, 41, 57
Spanish colonizers, 27–28
Spanish conquest, 33–35
Spanish control, 38–39
Spanish language, 53–54, 69
Statehood options, 45
Subsistence farming, 71
Sugar cane, 34, 71

Tainos, 32–33, 49–50
Taxes, property, 70
Tectonic plates, 26
Temperatures, 23–24
Tobacco, 71
Today in Puerto Rico, 81–87
  the arts, 82–85
  overview, 81
  social problems, 85–87
Tourism, 19, 44–45, 79–80, 88
Trade winds, 22–23
Tsunamis, 27

Unemployment levels, 87
United States and Puerto Rico.
  See Puerto Rico
Universidad Interamericana de Puerto Rico, 66
University of Puerto Rico, 64–65, 68
Urbanization, rapid, 72, 90
U.S. Internal Revenue Code, Section 936, 76

Index

U.S. occupation, 57–61
U.S. Postal Service, 80

Venezuela, 49
Villa, Anibal Acevedo, 89
Virgin Islands, 11

Water shortages, 89
Whites, 52

Women, voting rights of, 59–61
Women's Movement, Proletarian, 61
World War II, 61, 73

Yoruba communities, 51

Zones, climate, 23

Picture Credits

page:

- 10: New Millennium Images
- 12: © Lucidity Information Design
- 15: © Lucidity Information Design
- 20: New Millennium Images
- 25: KRT/NMI
- 30: New Millennium Images
- 36: New Millennium Images
- 41: © CORBIS
- 47: KRT/NMI
- 51: KRT/NMI
- 60: © Bettmann/CORBIS
- 65: New Millennium Images
- 69: AFP/NMI
- 74: Hulton Getty Photo Archive/NMI
- 79: New Millennium Images
- 84: New Millennium Images
- 89: EPA Photos/NMI

Cover: New Millennium Images

About the Contributors

JOSÉ JAVIER LÓPEZ is Associate Professor of Geography at Minnesota State University (M.S.U.), in Mankato. He was born in San Juan, Puerto Rico, and received a doctorate degree in Geography from Indiana State University. Following graduation, he moved to Minnesota and began a career in geographic education. Since becoming a faculty member of M.S.U.'s College of Social and Behavioral Sciences, José Javier López has researched various issues pertinent to the social geography of the United States and Latin America. Over the years, he has taught many systematic courses, including Spatial Statistics, Spatial Analysis, Cultural Geography, Economic Geography, Social Geography, and Rural Development. In addition to his teaching, José Javier López enjoys scuba diving in the Caribbean and traveling the diverse regions of the Americas.

CHARLES F. "FRITZ" GRITZNER is Distinguished Professor of Geography at South Dakota University in Brookings. He is now in his fifth decade of college teaching and research. During his career, he has taught more than 60 different courses, spanning the fields of physical, cultural, and regional geography. In addition to his teaching, he enjoys writing, working with teachers, and sharing his love for geography with students. As consulting editor for the MODERN WORLD NATIONS series, he has a wonderful opportunity to combine each of these "hobbies." Fritz has served as both President and Executive Director of the National Council for Geographic Education and has received the Council's highest honor, the George J. Miller Award for Distinguished Service. In March 2004, he won the Distinguished Teaching award from the Association of American Geographers at their annual meeting held in Philadelphia.